An Introduction to GaAs IC Design

An Introduction to GaAs IC Design

S. J. Harrold

Prentice Hall
New York London Toronto Sydney Tokyo Singapore

First published 1993 by
Prentice Hall International (UK) Ltd
Campus 400, Maylands Avenue
Hemel Hempstead
Hertfordshire, HP2 7EZ
A division of
Simon & Schuster International Group

Typeset in 10/12 pt Times
by MHL Typesetting Ltd, Coventry

Printed and bound in Great Britain by
Dotesios Ltd, Trowbridge, Wiltshire.

Library of Congress Cataloging-in-Publication Data

Harrold, S.J.
An introduction to GaAs IC design / S.J. Harrold.
p. cm.
ISBN 0-13-486358-5 (pbk.)
1. Integrated circuits — Design and construction.
2. Gallium arsenide semiconductors. I. Title.
TK7874.H3915 1993
621.3815′2–dc20 92-22641
CIP

British Library Cataloguing in Publication Data

A catalogue record for this book is available from the British Library

ISBN 0-13-486358-5 (pbk)

1 2 3 4 5 96 95 94 93 92

Contents

Acknowledgments viii
Abbreviations ix

Introduction xi

1 Properties of GaAs 1

1.1 Electron mobility 2
1.2 Semi-insulating substrate 5
1.3 Radiation hardness 7
1.4 Integrated opto-electronics 9
1.5 Noise properties 10
1.6 Hole mobility 11
1.7 Cost 11
1.8 Fragility 12
1.9 Thermal conductivity 12
1.10 Surface energy states 12

2 GaAs Device Operation 13

2.1 Bipolar transistors 13
2.2 IGFETs 15
2.3 JFETs 16
2.4 GaAs MESFETs 18
2.5 Analysis of GaAs circuits 31

3 Digital Design 34

3.1 Inverters 34
3.2 Direct-coupled FET logic (DCFL) 39
3.3 Buffered FET logic (BFL) 40
3.4 Schottky diode FET logic (SDFL) 42
3.5 Source-coupled FET logic (SCFL) 43
3.6 Alternative logic circuits 46
3.7 GaAs logic noise margins 49
3.8 Complex gates 55
3.9 Logic design techniques 61
3.10 Speed considerations 64
3.11 Input buffers 71
3.12 Output buffers 73

4 Analog Design 79

4.1 Circuit analysis and d.c. biasing 79
4.2 Analog building blocks 83
4.3 Design examples 105

5 Processing 110

5.1 Crystal growth 110
5.2 Active layer preparation 114
5.3 GaAs devices 117
5.4 Photolithography 118
5.5 Device isolation 120
5.6 Annealing 121
5.7 Contacts and interconnect 122
5.8 Passivation and dielectric layers 123
5.9 Threshold voltage determination 124
5.10 Reduction of parasitic source and drain resistance 126
5.11 Packaging 128

6 Layout and System Considerations 132

6.1 Layout rules 132
6.2 Layout considerations 135
6.3 System considerations 146

7 Future Developments **151**

7.1 High electron mobility transistors (HEMTs) 151
7.2 Heterojunction bipolar transistors (HJBTs) 153
7.3 Opto-electronic integration 155

References 159
Further reading 165
Index 167

Acknowledgments

The conception of this book arose from experiences gained in periods of employment at Plessey (Caswell) Research Ltd, STC Technology Ltd, and then UMIST. I would like to acknowledge the contributions of the many people I have worked with who have directly and indirectly helped to refine the ideas included in the book. Specifically, mention should be made of Alan Hughes (ex-Plessey), Ian Vance, Joe Mun, John Phillips and Graham Barker of STC, David Haigh of University College, London, and Chris Toumazou of Imperial College. Finally, I would like to thank the students of the Integrated Circuit System Design (ICSD) and VLSI Systems Engineering MSc courses at UMIST for their various comments on the set of course notes which formed the basis for this book. Thank-you.

Abbreviations

2DEG	Two-dimensional electron gas
a.c.	Alternating current
APD	Avalanche photo-diode
BFL	Buffered FET logic
CCL	Capacitor-coupled logic
CDFL	Capacitor-diode FET logic
CMOS	Complementary MOS circuits
CMRR	Common-mode rejection ratio
D-MESFET, DFET	Depletion-mode MESFET
d.c.	direct current
DCFL	Direct-coupled FET logic
DHJBT	Double heterojunction bipolar transistor
E-MESFET, EFET	Enhancement-mode FET logic
ECL	Emitter-coupled logic
FET	Field effect transistor
GaAs	Gallium arsenide
HB	Horizontal Bridgman
HEMT	High electron mobility transistor
HJBT, HBT	Heterojunction bipolar transistor
IC	Integrated circuit
IGFET	Insulated-gate field effect transistor
JFET	Junction field effect transistor
LEC	Liquid encapsulated Czochralski
LED	Light-emitting diode
LSI, VLSI	Large scale integration, Very LSI
MBE	Molecular beam epitaxy

MESFET	Metal–semiconductor FET
MIM capacitor	Metal–insulator–metal capacitor
MISFET	Metal–insulator–semiconductor FET
MOCVD	Metal–organic chemical vapour deposition
MOSFET	Metal–oxide–silicon FET
MQW	Multi-quantum well
NM	Noise margin
NMOS	*n*-channel MOS circuits
OEIC	Opto-electronic IC
PIN diode	P-Intrinsic-N diode
RAM	Random access memory
RC	Resistance–capacitance product
rf	radio frequency
SAINT	Self-aligned implantation for N^+ layer technology
SBFL	Super buffer FET logic
SCF	Switched-capacitor filter
SCFL	Source-coupled FET logic
SDFL	Schottky diode FET logic
TTL	Transistor–transistor logic
VPE	Vapour-phase epitaxy

Introduction

In the early days of the development of GaAs IC technology, silicon IC manufacturers were fond of saying that GaAs was 'the technology of the future, always has been . . . always will be'. Although this cynical view can be understood when one considers the problems associated with the processing of any new technology and also the phenomenal rate at which silicon ICs have improved in performance, GaAs ICs still offer considerable advantages. Today the technology problems have been overcome to the extent that many GaAs IC manufacturers now offer parts of high quality in very large quantities, and many instrument and systems builders use GaAs ICs to yield improved performance. It can truly be said that GaAs has come of age; it is no longer a technology which is restricted to the laboratory and research conferences.

Although the use of GaAs has proliferated, the number of engineers with experience in the design of GaAs ICs is still small. This book is intended to help remedy this situation. It is aimed to satisfy the needs of undergraduate students of electronic engineering or computer science, and also practising silicon IC designers. It is not anticipated that a reader of this book will instantly be able to design a world-beating IC, that would require a knowledge of process and design details of which there are too many variants specific to individual manufacturers to cover in an introductory text. However, this book will give an understanding of the general principles which can be applied to all GaAs IC designs. The book aims to give sufficient information so that a first design attempt can be made, and so that the designer and the manufacturer can interact and talk the same language, without the 'communications gap' present between existing silicon designers and GaAs IC manufacturers.

The contents of the book concentrate on ICs for digital and analog applications. Microwave circuits are specifically excluded: these circuits require specialized processing and design techniques which are already well described in other texts (see Further Reading). In the treatment of digital and analog ICs, the emphasis is on the design and analysis of circuit 'building blocks' which can be used to create complex integrated systems, and on the various trade-offs which exist between the various circuit properties. Techniques for using these building blocks vary little between GaAs and silicon, and are described in many other books. More advanced discussions on GaAs system design aspects may be found in the various texts listed in Further Reading.

The book is organized so as to reflect the order in which a newcomer to GaAs will need various information. To begin with, Chapter 1 reviews the properties of GaAs compared with silicon, and demonstrates why interest in GaAs devices and ICs has arisen. Chapter 2 then discusses the properties of GaAs devices, and gives parameters which may be used to simulate the behaviour of GaAs circuits. Chapters 3 and 4 concentrate on digital and analog circuits, respectively. It should be noted that at very high speeds digital circuits have analog characteristics and cannot be simply considered as sources of digital 1s and 0s, and there is thus a degree of cross-referencing between these two chapters. Chapter 5 then describes how GaAs devices and ICs are fabricated, and this is followed in Chapter 6 by a discussion of layout rules and various considerations which must be obeyed if an IC or system is to work reliably at high speed. This chapter also gives a set of generic layout rules which should allow a working IC to be fabricated by most GaAs IC manufacturers. Finally, Chapter 7 gives a brief description of various alternative devices. Although GaAs-based, these devices rely upon the epitaxial growth of various ternary compounds such as $Al_{(1-x)}Ga_xAs$, and this difficult growth process limits the application of these devices at present. This situation is not expected to remain static, and future editions of this and other books can be expected to have enlarged sections covering the use of these devices.

1 □ Properties of GaAs

The aim of this book is to describe how gallium arsenide (henceforth abbreviated to its chemical symbol GaAs) can be used to implement integrated circuits for digital and analog applications. Before looking into the details of how these integrated circuits (ICs) are designed, however, it is pertinent to ask the question 'Why GaAs?'. What is it that makes the use of GaAs so much more desirable than silicon, bearing in mind that the technology required to process silicon ICs has now matured to the point where extremely complex designs can be fabricated at remarkably low cost? The answer to this question lies in the electronic and physical properties of GaAs, which are compared with those of silicon in Table 1.1. Some of these properties have an advantageous effect on IC performance, while other properties produce a deleterious effect. A summary of the relative advantages and disadvantages is given in Table 1.2; the rest of this chapter is devoted to explaining the terms used in this table and illustrating in more detail how the various properties affect the performance of a GaAs IC. Subsequent chapters are devoted to describing the actual devices, processing, and design techniques needed to produce such an integrated circuit.

Table 1.1 Electronic and physical properties of GaAs and silicon

Property	GaAs	Si	Units
Electron mobility ($N_D = 10^{17}$ cm^{-3})	4000	800	cm^2 V^{-1} s^{-1}
Electron saturation velocity	1.4×10^7	6.5×10^6	cm s^{-1}
Hole mobility ($N_A = 10^{17}$ cm^{-3})	250	350	cm^2 V^{-1} s^{-1}
Dielectric constant	12.6	12	
Intrinsic resistivity	10^9	10^6	ohm cm
Energy gap	1.43 (direct)	1.12 (indirect)	eV
Schottky barrier height	0.7–0.8	0.4–0.6	V
Thermal conductivity	0.9	1.5	W cm^{-1} K^{-1}

Table 1.2 Comparison of GaAs and silicon

Advantages of GaAs vs Si	Disadvantages of GaAs vs Si
Higher electron mobility	Lower hole mobility
Lower noise at high frequencies	Higher noise at low frequencies
Semi-insulating substrate	High density of energy states at surface
Higher radiation hardness	Higher cost
Opto-electronic integration possible	Great fragility
	Lower thermal conductivity

1.1 Electron mobility

The resistivity of a doped semiconductor is dependent upon the number of charge carriers (holes or electrons) present within the material (i.e. the doping density) and also upon the ease with which these carriers can move under the application of an electric field. This latter property is assessed by the mobility of the carriers, which is defined as the ratio of the carrier velocity to the electric field strength. GaAs has an electron mobility which is of the order of five times higher than silicon, and the effect of this is that devices fabricated in GaAs will have a much lower resistivity than silicon devices of similar doping density. An alternative way of looking at this is to say that GaAs devices of the same resistance as silicon equivalents can be achieved with much smaller (narrower) dimensions.[1] The higher electron mobility of GaAs has a direct impact on the speed and power dissipation of an electronic circuit due to this reduced resistivity.

If we consider two functionally identical electronic systems, one built using *n*-type GaAs and the other using *n*-type silicon (i.e. electrons are the charge carriers), then if we require both systems to have the same power dissipation the GaAs system must use narrower devices due to the lower resistivity. If the GaAs devices are smaller than the silicon equivalents, then the associated capacitance will also be smaller (the dielectric constants of GaAs and silicon are approximately equal). This lower capacitance affects the speed of the system since this speed is dependent upon the rate at which the signals within the system can change. To a good approximation the rate of change of the signals is determined by the RC time constants formed by the product of the output impedance of the sources and the capacitive load on these outputs (see Figure 1.1). The output impedance is determined by the amount of current which can be sunk or sourced, and this is related to the resistances present; at similar power dissipations the output impedances present in GaAs and silicon systems will thus be similar. The lower capacitance present in the GaAs system thus produces a faster system.

By using devices with a higher resistance (i.e. narrower or longer) it is possible to design a GaAs system to run at the same speed as a silicon system, but in this case at a reduced current and thus reduced power dissipation. A trade-off exists between the speed and the power of a system. Digital IC technologies are usually compared using a 'power-delay

1. Dimensions perpendicular to the direction of current flow are referred to as 'widths' (or 'areas'), while dimensions parallel with the current flow are referred to as 'lengths'. The current is proportional to the width or area, and inversely proportional to the length of the device.

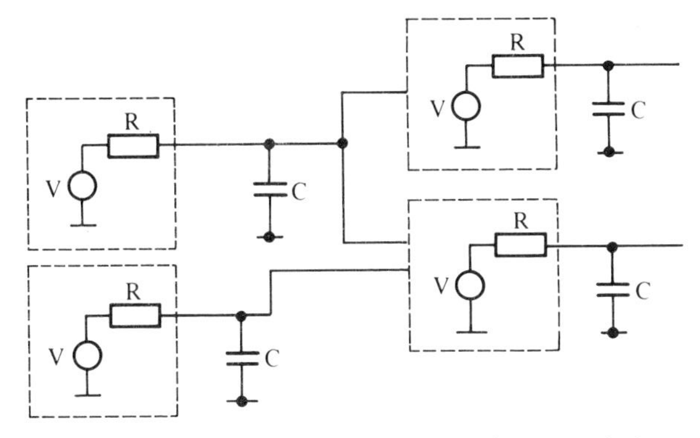

Figure 1.1 'Black-box' view of a section of a general electronic system

product' parameter, which is calculated by multiplying the logic gate delay by the power dissipation per gate. A high power-delay product implies that either the gate power dissipation is high (limiting the integration density possible from that technology), or that the gate delays are high (limiting the speed performance of the technology), or both. The aim of all integrated circuit processing engineers is to minimize the power-delay product of their technology so as to improve both the level of integration possible and the speed. GaAs has a power-delay product advantage over silicon, which is principally determined by the ratio of electron mobilities. Ideally, this is approximately 5 : 1; in practice, the performance of any system will be degraded by circuit parasitics such as stray capacitance and resistance. A different processing technology and the relative immaturity of this processing for GaAs means that the power-delay product advantage is reduced by these parasitics to approximately 3 : 1.

Examples of this performance advantage are shown in Table 1.3, which lists the

Table 1.3 Examples of commercially available ICs

GaAs IC	Equivalent silicon IC
Vittesse Electronics VE29G10A Bit-slice microcontroller 100 MHz clock 1.7 W	Advanced Micro Devices 2910A Bit-slice microcontroller 10 MHz 1.3 W
Gigabit Logic 10G021A Dual D-type DC to 2.7 GHz 600 mW	GEC-Plessey SP9131 Dual D-type DC to 520 MHz 360 mW
Anadigics AOP1510 op-amp 150 MHz gain-bandwidth product 7.5×10^8 V s^{-1} slew-rate 0.9 W	Harris HA2541 op-amp 40 MHz gain-bandwidth product 2.8×10^8 V s^{-1} slew-rate 1.5 W

performance of some commercially available GaAs ICs and their equivalent silicon components. If the powers and speeds are compared it should be clear that the GaAs ICs exhibit a useful advantage. In the case of the bit-slice microcontroller, this advantage is 7.5 : 1, which reflects the older technology used for the Advanced Micro Devices silicon version; however, generally this advantage is approximately 3 : 1 reflecting the electron mobility advantage of GaAs as described earlier.

Table 1.4 compares the performance of various silicon and GaAs gate arrays (details concerning the circuitry of the different GaAs logic families are given in Chapter 4). The performance of all these arrays is shown plotted in Figure 1.2. Again it can be seen that GaAs exhibits a power-delay product advantage over silicon of approximately 3 : 1. This advantage is also shown in Figure 1.3 which compares the performance of typical GaAs and silicon static random access memories.

For state-of-the-art (high performance) applications this advantage is worth fighting for. As time proceeds it can be expected that silicon technology will improve to the point where it can take over from the present GaAs circuits to give substantial cost and yield improvements. However, we can also expect GaAs technology to improve correspondingly, thus maintaining its advantage for high performance applications. Early reports on GaAs ICs suggested that the high speed property might enable GaAs to replace the use of silicon. This is no longer suggested as a variety of other reasons (chiefly cost) make the continued use and development of silicon highly desirable. In fact, silicon devotees now suggest that further technology improvements will eventually allow silicon to replace GaAs! The truth probably lies between these two extremes. There will always be a place for both GaAs and silicon, silicon having the bulk of the market. However, where high performance is essential and cost is not the limiting factor, then GaAs is able to provide a level of performance unobtainable by any other means.

Table 1.4 Comparison of commercial gate arrays fabricated in silicon and GaAs

Gate array	Gate delay (speed)	Gate power	Power-delay product (fJ)
Silicon			
Hughes U-series (2 micron CMOS)	400 ps	8 W MHz^{-1} = 1 mW @ 125 MHz	400 fJ
Thomson ECL/TTL	0.8 ns	0.1 mW	80 fJ
Nippon bipolar	500 ps	1 mW	500 fJ
NTT bipolar	78 ps	2.6 mW	203 fJ
GEC–Plessey DS-series (bipolar)	1 ns	0.132 mW	132 fJ
Siemens bipolar	150 ps	7.5 mW	1125 fJ
Toshiba CMOS/SOS	870 ps	0.45 mW	392 fJ
NEC ECL4	100 ps	3.7 mW	370 fJ
Gallium arsenide			
OKI SBFL	390 ps	0.27 mW	105 fJ
Toshiba DCFL	215 ps	0.5 mW	108 fJ
Tektronix DCFL	200 ps	0.25 mW	50 fJ
NEC BFL	50 ps	3.5 mW	175 fJ

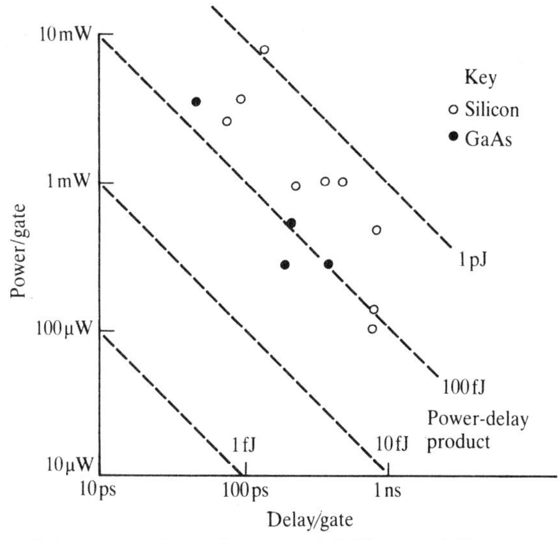

Figure 1.2 Comparison of commercial silicon and GaAs gate arrays (see Table 1.4)

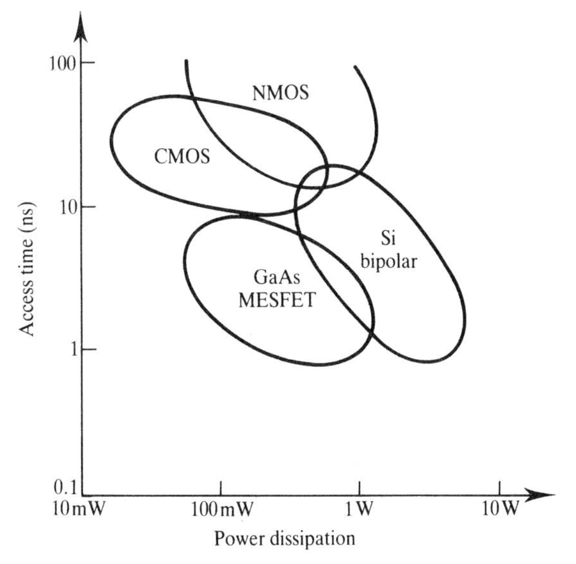

Figure 1.3 GaAs and silicon static RAM performance

1.2 Semi-insulating substrate

When electronic components are fabricated together as an integrated circuit it is important that any interaction between the components which might affect the performance of the complete circuit is kept to a minimum. The integrated components must thus be electrically isolated from each other. The resistivity of a semi-conductor is dependent upon the number of charge carriers present, and undoped (intrinsic) silicon has a resistivity which is typically of the order of 10^6 ohm cm. This resistivity value means that good isolation will only be attained if devices are widely separated; it is not high enough to provide sufficient isolation in an integrated circuit, where the aim is to pack devices close together in order to achieve high device counts per chip. Silicon technologies therefore generally use a doped n-type or p-type substrate with devices fabricated in oppositely doped 'wells' (see Figure 1.4). Any two devices will then always be separated by two back-to-back $p-n$ junction diodes, and this provides sufficient device isolation at relatively small separations. In comparison, the intrinsic resistivity of GaAs is several orders of magnitude higher, falling into what is sometimes called the 'semi-insulating' range. This property allows GaAs devices to be fabricated in wells in an undoped substrate while still maintaining good isolation (see Figure 1.5). The principal advantage this offers is to reduce the parasitic capacitance in GaAs ICs.

In silicon, the principal parasitic capacitance component is to the back of the wafer (usually held at ground potential), and is associated with the device wells and the interconnecting tracks on top of the substrate (see Figure 1.6). The parasitic capacitance of the device well is determined by the well size and the thickness of the depletion layer associated with the well-substrate $p-n$ junction. The track parasitic capacitance is determined by the track dimensions and the thickness of the oxide layer which separates the tracks from the doped substrate. The depletion layer thickness and the oxide thickness are typically measured in tenths of a micron, which results in parasitic capacitances roughly of the order of 1 fF μm^2. In comparison, the availability of a semi-insulating substrate in GaAs ICs means that the well and track parasitic capacitances in a GaAs IC are determined by the well size (which may be smaller due to the higher electron mobility as noted in Section 1.1), the track dimensions (which will be similar to those on a silicon

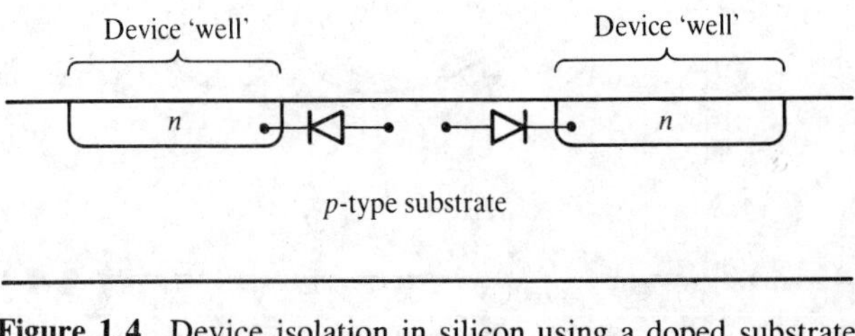

Figure 1.4 Device isolation in silicon using a doped substrate

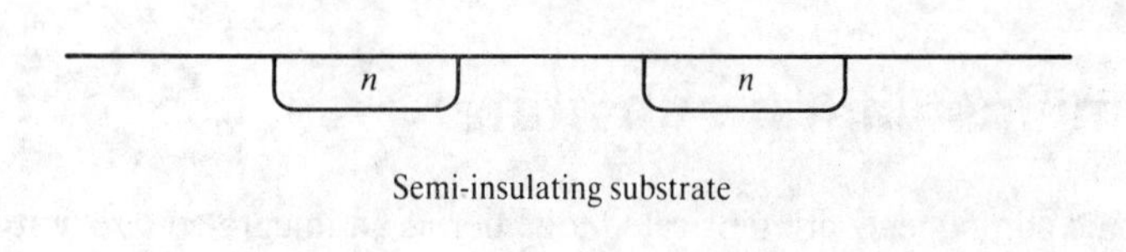

Figure 1.5 Device isolation in GaAs using a semi-insulating substrate

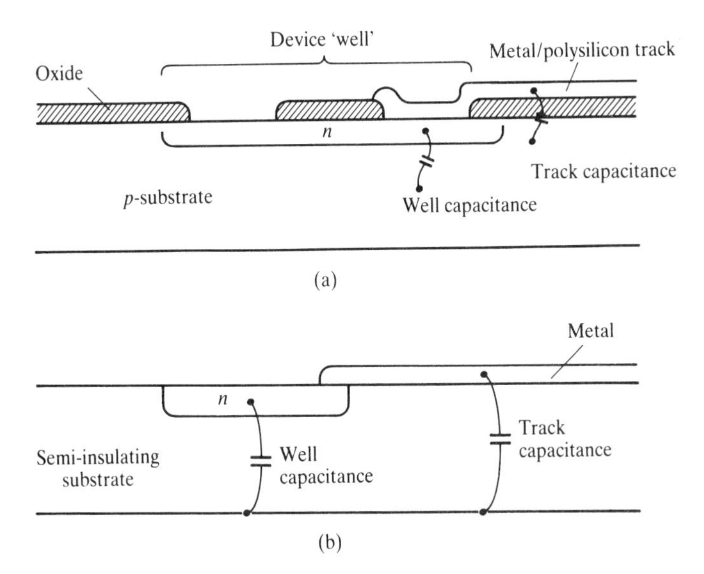

Figure 1.6 Principal parasitic capacitances in silicon and GaAs ICs: (a) silicon; and (b) GaAs

IC) and the thickness of the semi-insulating wafer (see Figure 1.6). The wafer thickness is typically 350 μm, i.e. approximately three orders of magnitude greater than the silicon oxide and the well depletion-layer thicknesses. The parasitic capacitance to ground is thus reduced by a corresponding factor. (In many GaAs ICs the parasitic capacitance to ground is so small that track-to-track capacitance forms the dominant parasitic component.)

The level of parasitic capacitance is important since in most practical IC technologies, it is the parasitic capacitances rather than the intrinsic device capacitances which limit the extent to which IC speeds can be increased. If the parasitic capacitances can be reduced, then it becomes possible to operate ICs at much higher frequencies. The semi-insulating substrate used in GaAs ICs thus contributes to the speed advantage of GaAs over silicon. As a point of interest, it should be noted that silicon IC technologies do exist (e.g. silicon-on-sapphire (SOS) or silicon-on-insulator (SOI) technologies [1]) which also use a semi-insulating substrate and which also give a speed advantage over standard silicon IC technologies, although without the added benefit of a high electron mobility.

1.3 Radiation hardness

The immunity of an IC to damage by exposure to radiation is important for many applications. Systems for military, space and nuclear industry have particular requirements for radiation hardness, but other applications may also be adversely affected if an IC is sensitive to radiation naturally occurring from sources in the environment or packaging. GaAs has a variable advantage over silicon, dependent upon the type of radiation concerned. This sensitivity is described below.

1.3.1 Total dose immunity

Radiation can ionize the atoms in a material, with the total ionized charge produced being dependent upon the total radiation dose. Silicon ICs are sensitive to total dose effects through charge build-up in the surface oxide film; this charge can alter the threshold voltage and parasitic resistances of metal–oxide–silicon (MOS) devices and allow large parasitic surface leakage currents in bipolar devices. Typically silicon MOS ICs fail at approximately 10^4 rads (although with special processing this may be increased to approximately 10^6 rads), while silicon bipolar ICs are less sensitive and fail at approximately 10^5 rads. In comparison GaAs ICs are relatively immune to total dose effects from ionizing radiation and only undergo a small change in their operating parameters (e.g. the gate propagation delay) at levels as high as 5×10^7 rads [2]. It is thought that this immunity is related to the inability to produce IGFET-type structures in GaAs (see Chapter 2). GaAs has a high density of energy states at the surface which 'soak-up' any charge produced by the radiation and thus prevent any changes in the device parameters (such as threshold voltage or parasitic resistance) and also prevent large surface leakage currents from occurring.

1.3.2 Transient radiation

One of the effects of radiation is to generate extra hole–electron pairs in the irradiated material. These charge carriers may move under the influence of any electric field present to affect the device characteristics. In the absence of the radiation dose, recombination will eventually remove all the excess hole–electron pairs giving a return to the original characteristics. The sensitivity of the device characteristics to change during the period of the radiation pulse is thus referred to as the transient radiation hardness.

The transient radiation currents may occur in two regions: within the active devices or in the substrate. In GaAs ICs the dominant effects are in the substrate [3]. The wider bandgap of GaAs over that of silicon makes it harder to generate carriers and thus the sensitivity of GaAs to transient radiation is relatively low provided high quality, undoped substrates are used. It may be noted that a correspondingly high immunity to transient radiation is also given by silicon-on-sapphire structures, which also use a wide bandgap substrate [4].

Table 1.5 compares the transient radiation immunity of various GaAs and silicon

Table 1.5 Immunity of GaAs and silicon memories to transient radiation pulses

Technology	Transient radiation immunity
Analog circuits and sensors	5×10^5
Static NMOS	2×10^6
CMOS	3×10^6
Bipolar	3×10^6
'Hardened' TTL	10^7
GaAs D-MESFET	10^7
GaAs E/D JFETs	10^8

memories [5]. It can be seen that the best immunity is given by GaAs JFET circuits. It is thought that these JFETs have a better immunity than GaAs MESFETs due to the increased distance of the active region from the surface. (The structures of the GaAs MESFET and JFET referred to in this Table are described in Chapter 2.)

1.3.3 'Soft-errors'

Soft-errors are random single-event errors caused by sensitivity to cosmic rays, high energy protons or alpha-particles (from, for example, packaging materials). Sensitivity to these soft-errors is the one aspect of radiation hardness where GaAs at present does not give a significant advantage over silicon. The first tests on GaAs RAMs showed a similar sensitivity to silicon *n*-channel MOS (NMOS) circuits: better than silicon bipolar memory but many orders of magnitude worse than complementary MOS (CMOS) circuits [2]. Later GaAs designs with complementary JFETs promise similar immunity to standard CMOS RAMs [6].

Whether soft-errors are a problem or not depends a great deal on the system and the application. For instance, since these errors are random single events, error-detection and correction circuitry on board each RAM may greatly reduce the sensitivity. Alternatively, different circuit designs and device structures may be found which confer an enhanced radiation hardness. It can be expected that improvements can be gained if further attention is paid to the specific development of radiation-hard GaAs ICs.

1.4 Integrated opto-electronics

At very high switching speeds, one of the biggest problems concerning electronic systems is chip-to-chip communication. As switching speed increases, so does the size of the output buffers necessary to provide fast voltage changes, and the chip power dissipation increases correspondingly. Furthermore, problems with maintaining good signal integrity on the interconnecting lines become more acute at high switching speeds; any mismatch between the line impedance and the line load can cause severe 'ringing' on the signal edges (see Figure 1.7). All signal lines have to be treated as transmission lines at very high frequencies,

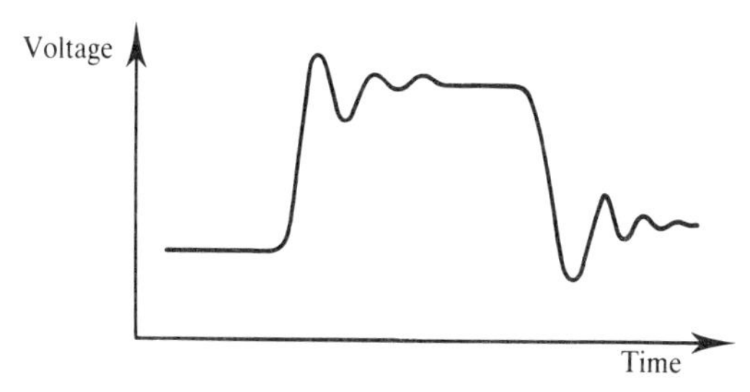

Figure 1.7 'Ringing' caused by mismatch between signal line characteristic impedance and load

and this greatly complicates the process of printed circuit board (p.c.b.) design and production, particularly where multiple chips are being driven from one signal line. This problem is discussed further in Chapter 6.

One possible solution to this problem is to use optical communication between chips, integrating a laser or light-emitting diode (LED) and a photodetector on chip. Although the processing involved is complicated, GaAs can be used to integrate electronic functions and also optical circuits (using AlGaAs/GaAs structures) on the same chip. GaAs thus has a good advantage in this respect over silicon, where the lattice mismatch between silicon (used for electronic circuits) and suitable light-emitting materials (e.g. AlGaAs) sets up a strain in the device structures which makes the manufacture of good laser/LED devices extremely difficult.

1.5 Noise properties

Electrical noise in GaAs devices (which are normally FET structures as described further in Chapter 2) generally arises from three sources. These are:

(a) Thermal noise in the channel resistance and also the source, drain and gate parasitic resistances. The noise power is dependent upon the resistor value and is independent of frequency.
(b) Thermal noise in the channel which is capacitively coupled to the gate and then amplified by the FET. The capacitive coupling gives a noise power which is proportional to frequency.
(c) Flicker (or $1/f$) noise from the random generation/recombination of carriers in energy states associated with defects in the material [7, 8]. The noise power from this source obeys a $(1/f)^n$ law, where n is close to unity.

The total noise power from a GaAs FET is built up from these three sources as shown in Figure 1.8. Good device design and processing can minimize the channel and gate thermal noise by minimizing the series resistance and the gate capacitance. Comparing GaAs and

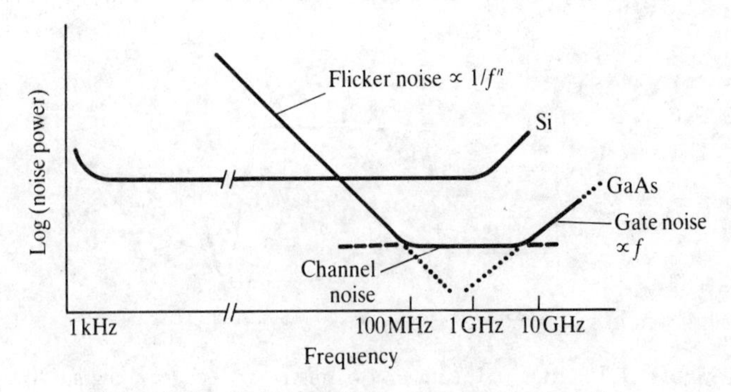

Figure 1.8 GaAs and silicon noise characteristics

silicon, the high electron mobility of GaAs and the correspondingly low resistivity mean that the thermal noise power generated in GaAs FETs is very low. GaAs is thus very attractive for use in low noise amplifiers at microwave frequencies. Unfortunately, the corner frequency at which flicker noise becomes comparable with the thermal noise is very much higher with GaAs (typically 100 MHz compared to kilohertz frequencies for silicon) and so GaAs is not as attractive for applications requiring low noise amplification at low frequencies. Systems requiring low noise over a very wide bandwidth from (say) a few hertz to many gigahertz would require a combined silicon–GaAs system. Comparison of the typical noise characteristics of GaAs and silicon is also shown in Figure 1.8.

1.6 Hole mobility

The hole mobility of GaAs is lower than that of silicon by a factor of approximately 5 : 7. By following the same line of reasoning used for considering the effect of a high electron mobility, it should be clear that devices based upon the use of *p*-type GaAs will be slower than those based upon *p*-type silicon. This applies whether one is considering bipolar devices or FETs, as is explained further in Chapter 2. *Npn* GaAs bipolar devices will have a higher parasitic base resistance, while *pnp* devices will have a longer base transit time and a higher parasitic emitter resistance. *P*-channel GaAs FETs will have a higher channel resistance. These parameters will degrade the speed in comparison to silicon.

In general then, devices requiring the use of *p*-type GaAs are not usually required. The exception to this is when a property other than speed is required. For example, it was mentioned in Section 1.3 that good soft-error radiation immunity can be obtained by using complementary JFET circuits (i.e. circuits using both *p*-channel and *n*-channel JFETs); this immunity cannot presently be obtained using *n*-type material only.

1.7 Cost

The major disadvantage of GaAs compared to silicon is the cost factor. This high cost occurs for several reasons. Not only are material costs higher (GaAs crystals are difficult to grow and IC production uses a lot of gold-based metallizations), but being a state-of-the-art technology means that processing is generally carried out by highly skilled engineers and technicians with correspondingly high labour costs. Capital costs are also high, for example, submicron gate lengths are now common necessitating the use of very expensive state-of-the-art photolithography equipment.

The biggest cause of the relatively high cost of GaAs ICs is the fact that generally yields are low and wafer sizes are relatively small compared to silicon. This means that there are fewer working devices per wafer over which the processing costs can be shared; i.e. the cost per device will be high. In practice, this means that the use of GaAs is restricted to high speed applications, where the high cost factor can be tolerated in order to obtain a degree of performance that is not available from silicon ICs.

1.8 Fragility

One of the most effective means of increasing wafer yields is to fabricate ICs using a standard, automated, process line with little human interaction. For example, consistent track widths can only be obtained if the photolithography exposure and development steps are reliably reproduced. Automated equipment is commonly used on silicon IC production lines, but it is difficult to use this standard equipment for processing GaAs due to its greater fragility — ideally, processed wafers should come out of the end of the line, not handfuls of dust! It is possible to use thicker substrates to offset the fragility, but this step imposes problems of its own (particularly those concerned with heat dissipation, as described in the next section). Generally, much greater care must be taken when handling GaAs compared to silicon.

1.9 Thermal conductivity

Any operating IC will dissipate heat, and a low thermal conductivity makes the problem of getting rid of this heat more severe. This problem can be reduced to some extent by thinning the substrate to reduce the thermal resistance, but if this is performed before processing the problem of fragility becomes more acute; if performed after processing then great care must be taken not to damage the fabricated IC. GaAs has a lower thermal conductivity than silicon, so that if substrate thinning is not carried out, then not only will GaAs ICs run hotter than a silicon IC dissipating the same power, but thermal gradients across the surface of the chip will be more severe. This latter effect may make it more difficult to equalize device parameters across the chip (e.g. gate delays or amplifier offset voltages), and may thus greatly affect the IC performance. When designing an IC layout, careful consideration should be given to the effect of any possible temperature gradients.

1.10 Surface energy states

Although GaAs possesses a thin native oxide layer, this oxide cannot be used in device fabrication to produce MOS-type circuits. A high density of energy states exists at the GaAs–oxide interface which prevents the operation of insulated-gate field effect transistor (IGFET) devices with useful characteristics. This is discussed further in Chapter 2.

The lack of IGFET devices means that the wealth of experience available from designing silicon NMOS and CMOS ICs cannot be directly exploited, and different circuit structures have to be derived. In partial compensation though, it should be noted that it is this same high density of surface states which confers the high immunity to total radiation dose effects.

2 □ GaAs device operation

When designing any electronic circuit (integrated or otherwise), there are generally three types of transistor from which we can choose to give the circuit the desired characteristics. These are:

(a) bipolar transistors (*npn* or *pnp*);
(b) insulated-gate field effect transistors (IGFETs); and
(c) junction field effect transistors (JFETs).

The following sections describe the properties of each of these types of transistor for a GaAs IC technology.

2.1 Bipolar transistors

Figure 2.1 shows a cross-section through the structure of a bipolar transistor; it can be seen that the transistor is constructed from a 'sandwich' formed by either $p-n-p$ doped material or $n-p-n$ doped material. The central region forms the emitter region, the base surrounds this and the collector is around the base. Under normal operating conditions the base-emitter $p-n$ junction is forward-biased and the base-collector region reverse-biased. At the base-emitter junction, charge carriers flow from the base into the emitter and also from the emitter to the base; the emitter is usually more highly doped so that the emitter-to-base current dominates. If the collector contact was left open-circuit, these charge carriers from the emitter would all flow to the base contact; however, the bias on the base-collector junction sets up a large electric field which sweeps the majority of the carriers into the collector region so that only a small percentage reach the base contact. Modulating the base current (i.e. the number of carriers flowing from the base to the emitter) will modulate the number of carriers injected by the emitter, and thus also the collector current. Since the collector current is much larger than the base current, the transistor thus offers current gain. It is this property which forms the basis of operation of all bipolar electronic circuits.

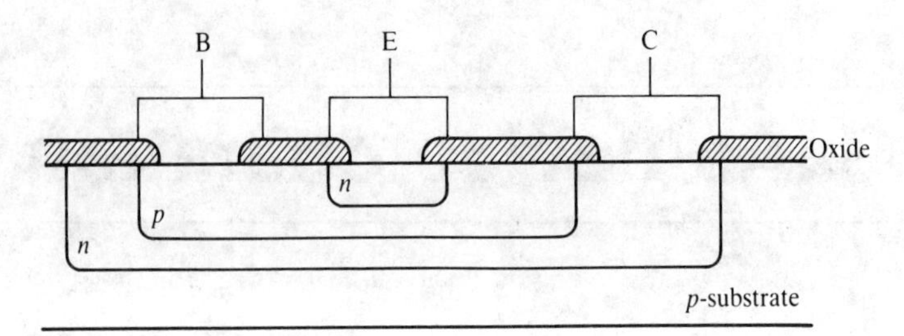

Figure 2.1 Cross-section through an integrated *npn* bipolar transistor

The speed of a bipolar transistor is determined by the speed with which the internal charge can be modulated. Charge is stored within the transistor in capacitance associated with the base-emitter and base-collector junctions; this charge cannot be modulated instantaneously, only at a rate determined by the RC time constants of the junction capacitances and the parasitic resistances through which the capacitances must be charged or discharged. These parasitic resistances arise from the physical separation of the transistor contacts from the junctions within it; Figure 2.2 shows their location within a typical bipolar transistor.

At the base-collector junction the capacitance is associated with the $p-n$ depletion layer. This capacitance is dependent upon the junction bias, the doping densities of the p and n-type regions, and the semiconductor dielectric constant. Thus at similar biasing and doping levels, GaAs and silicon bipolar transistors will have similar base-collector capacitances since the dielectric constants are very similar. The same argument applies to the base-emitter depletion layer capacitance; however, at this junction there is an additional capacitance associated with the accumulation, in the base region, of carriers which have been injected from the emitter, and which are in the process of being swept to the base-collector junction. This latter capacitance is dependent upon the transit time which the carriers take to cross the base region, and this in turn is dependent upon the carrier mobility. GaAs and silicon devices will thus have different base-emitter capacitances, and these will be dependent upon whether the injected carriers in the base are holes (*pnp* transistor) or

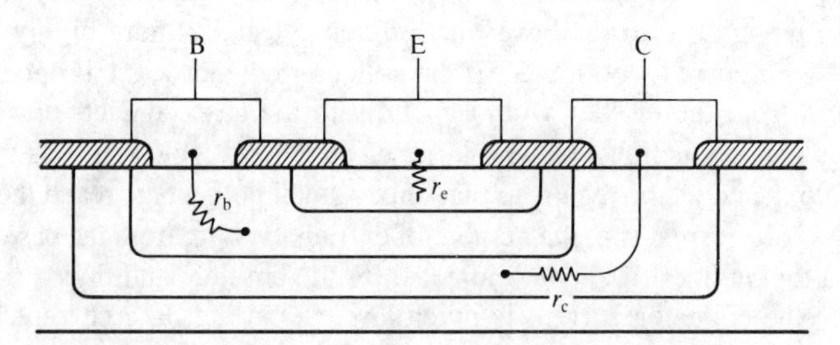

Figure 2.2 Location of parasitic resistances in a bipolar transistor

electrons (*npn* transistor). In *npn* structures, silicon will have the greater base-emitter capacitance, while in *pnp* structures GaAs will have the greater base-emitter capacitance.

Just as the base-emitter capacitance is mobility-dependent, so too are the parasitic resistances. Thus, comparing devices of similar dimensions, a GaAs *pnp* transistor will also have a greater emitter resistance than a silicon device, while a GaAs *npn* transistor will have a greater base resistance. Although the higher electron mobility of GaAs will reduce other parasitic resistances, it is the effect of the lower hole mobility which generally dominates the overall frequency response in both *npn* and *pnp* structures. GaAs bipolar transistors do not offer any speed advantage over silicon equivalents unless recourse can be made to special processing techniques to produce heterojunction bipolar transistors (HJBTs); these transistors are described further in Chapter 7.

2.2 IGFETs

Figure 2.3 shows the structure of an *n*-channel insulated-gate field effect transistor (IGFET). This is commonly known as a metal–oxide–silicon FET (MOSFET) when implemented in silicon technology. A metal **gate** is deposited over a *p*-type semiconducting substrate, separated from it by a thin insulating layer. Two $n+$ contacts (the **source** and **drain**) are provided to the substrate on opposite sides of the gate electrode — initially these contacts are electrically isolated from each other by the series back-to-back $p-n$ junctions formed between the $n+$ contacts and the *p*-type substrate.

The operation of the *n*-channel IGFET is illustrated in Figure 2.4. The operation of a *p*-channel device is identical, but with the polarity of all the dopants and the biasing reversed. As an increasing positive bias is applied to the gate, so an increasing negative charge is induced in the semiconductor. This charge initially resides in a depletion layer close to the insulator, but at large values of gate bias, charge is induced in a so-called inversion layer at the interface between the semiconductor and the insulator. In the inversion layer, the semiconductor effectively changes from *p*-type to *n*-type to create an *n*-channel which electrically connects the $n+$ source and drain contacts. A source–drain current may then flow in the channel if a potential difference is applied across the channel. The

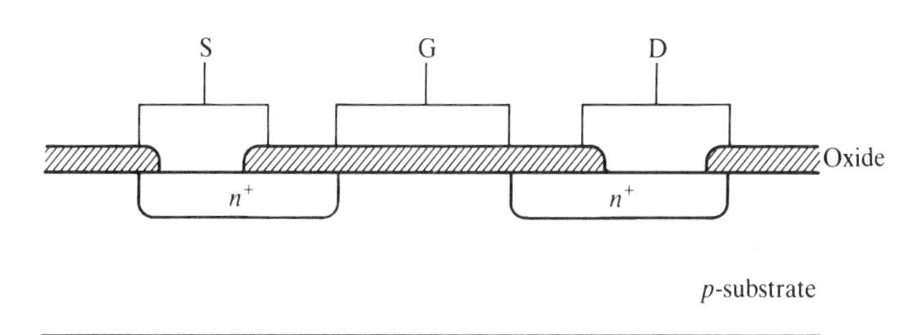

Figure 2.3 Cross-section through an integrated insulated-gate field effect transistor (IGFET): S, source; G, gate; D, drain

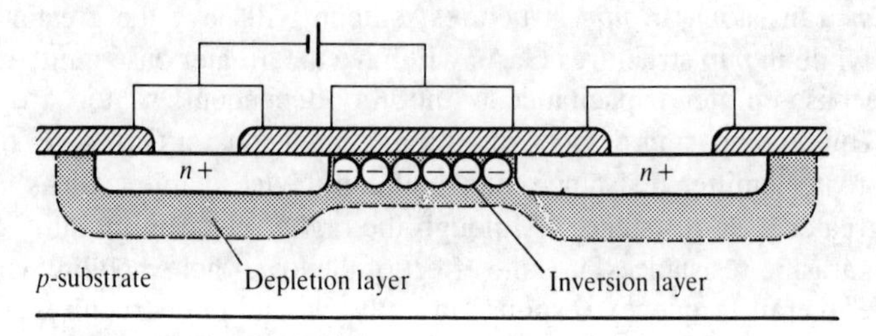

Figure 2.4 Formation of an inversion layer in an IGFET

resistivity of the channel is dependent upon the induced charge present in the inversion layer, so it can therefore be controlled by the gate bias; this control action forms the basis of operation of FET electronic circuits.

To obtain useful characteristics, it is important that the inversion layer can be created at a pre-defined, reproducible, gate bias — this parameter being known as the FET threshold voltage. Many years of research and development have been devoted to achieving this aim in silicon devices. The threshold voltage is very dependent upon the properties of the insulator–semiconductor interface; fortunately for the electronics industry, the Si–SiO_2 interface has the required desirable properties. In GaAs however, the situation is not so favourable. The interface between GaAs and the native oxide which forms on its surface on exposure to air has properties such that a large density of energy states is created at the interface. If a bias is applied to the gate on a GaAs IGFET, an opposite charge will be induced in the semiconductor, but it will be trapped in the interface states which thus prevent the modulation of the channel current by the gate bias. If a sufficiently large gate bias is applied, it may be possible to fill all the interface states so that the channel resistivity can then be modulated. However, the density of states is often so high that the breakdown voltage of the oxide may be exceeded before this is achieved. Much research has been devoted to finding suitable insulators giving a low density of states at the GaAs–insulator interface, but with little practical success [9]. Since the interface states are not able to trap charge instantaneously, it is possible to modulate the channel resistivity in a GaAs IGFET by applying the gate bias at a frequency greater than the capture rate associated with the states. This has been demonstrated using a silicon nitride insulator [10]. Unfortunately, this approach seems to have little practical application, since it is difficult to guarantee that the gate bias will always be modulated at frequencies greater than this minimum rate unless the signals are suitably encoded (which increases the circuit complexity) and also since it is very difficult to define the threshold voltage by this method. GaAs IGFETs are thus not a practical option with which to implement electronic circuits.

2.3 JFETs

An alternative form of the field effect transistor is the junction field effect transistor (JFET) whose structure is shown in Figure 2.5. JFET devices do not rely upon the formation

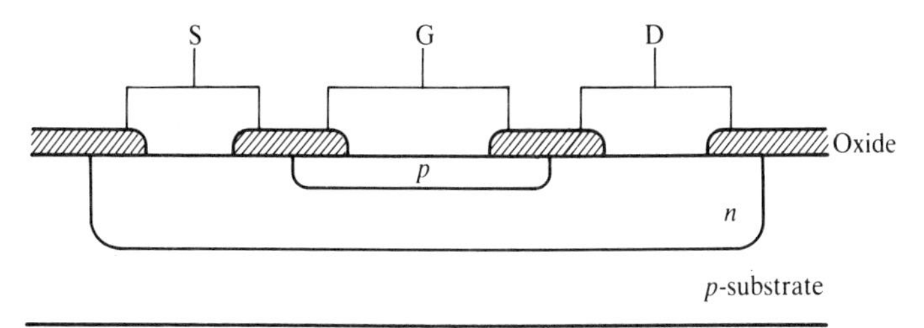

Figure 2.5 Cross-section through a junction field effect transistor (JFET): S, source; G, gate; D, drain

of an inversion layer for their operation, but instead the thickness (and hence the resistance) of a conducting channel is modulated by varying the thickness of a depletion layer above and/or below this layer (see Figure 2.6). The energy states associated with the GaAs surface do not prevent the modulation of the depletion layer width in this case, since the surface is in direct contact with a conductor which can easily supply sufficient charge to fill and empty these states.

The conventional JFET structure makes use of the depletion layer formed by a *p–n* junction; Figure 2.6 shows an *n*-channel device where the current in the FET is confined to a relatively thin *n*-type region under the gate electrode. The lower interface of the conducting channel can be defined either by using a *p*-type substrate so as to create a reverse-biased *p–n* junction at the substrate–channel interface, or (more common in GaAs) by using a semi-insulating (undoped) substrate. As with the bipolar and IGFET structures, the low hole mobility reduces the speed advantage of a GaAs JFET compared to a silicon equivalent, since changes in the channel thickness can only occur by modulating the depletion layer thickness and charge, which in an *n*-channel device entails moving holes through the *p*-type region between the gate contact and the channel. The relatively low hole mobility of a GaAs device affects the rate at which it can respond to changes in the gate bias. This could be compensated for by increasing the doping level in the *p*-type region; however, this would be accompanied by a corresponding increase in the gate capacitance.

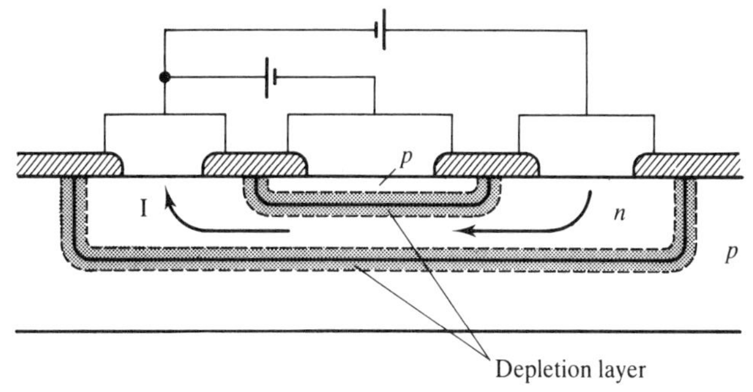

Figure 2.6 Formation of the conducting channel in a JFET (I = channel current)

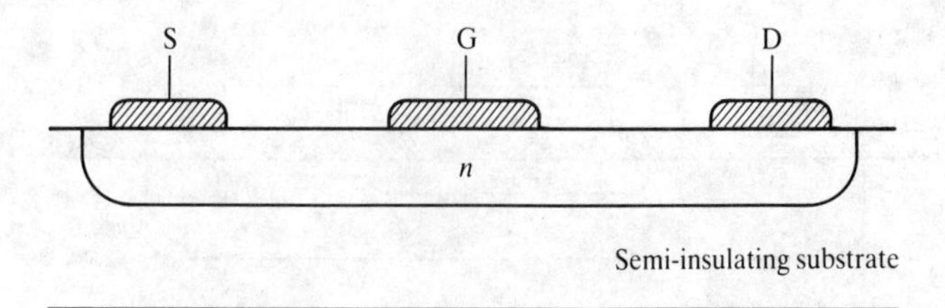

Figure 2.7 Cross-section through a metal–semiconductor field effect transistor (MESFET): S, source; G, gate; D, drain

In a GaAs *p*-channel device, the conducting channel will have a higher resistivity than in an equivalent silicon device, and again the circuit speed will suffer. Although it is possible to fabricate both *p* and *n*-channel JFET devices they are not generally used except where some other desirable property such as radiation hardness is required [6].

Although conventional GaAs JFETs offer little speed advantage compared to silicon, there is an alternative structure which does offer significant advantage. This structure relies upon the use of a metal–semiconductor Schottky diode instead of a *p–n* junction diode to modulate the channel resistivity. Figure 2.7 shows this structure which is known as the metal–semiconductor FET, or MESFET. In the *n*-channel device, no holes are required to either modulate the depletion layer thickness or to carry current in the channel. Full advantage can thus be taken of the relatively high electron mobility of GaAs compared to silicon to gain a significant speed advantage. Most ICs and discrete devices fabricated in GaAs make use of this *n*-channel MESFET device, whether for microwave or for high speed digital and analog applications. The following sections describe its characteristics in more detail.

2.4 GaAs MESFETs

2.4.1 Ungated devices (resistors)

If the Schottky contact is omitted, then at small drain–source voltages (V_{DS}) the channel current (I_{DS}) in a GaAs MESFET (i.e. a resistor) is linearly dependent on V_{DS} as given by:

$$I_{DS} = \frac{V_{DS}}{R_{DS}}$$

where R_{DS} is the total drain–source resistance and equals $2(r_c + r_{channel})$, r_c is the contact resistance and

$$r_{channel} = \frac{L_{DS}}{qn\mu_n aW}$$

where n = carrier density (i.e. doping density);
μ_n = low field drift mobility of carriers;

a = channel thickness (excluding surface depletion layer);
W = device width; and
L_{DS} = drain–source separation. (see Figure 2.8)

At larger V_{DS} this linear relationship breaks down to reflect the dependence of the electron velocity on electric field strength. Figure 2.9 illustrates this dependence for silicon and GaAs [11]. Whereas silicon exhibits a steady increase in electron velocity to field

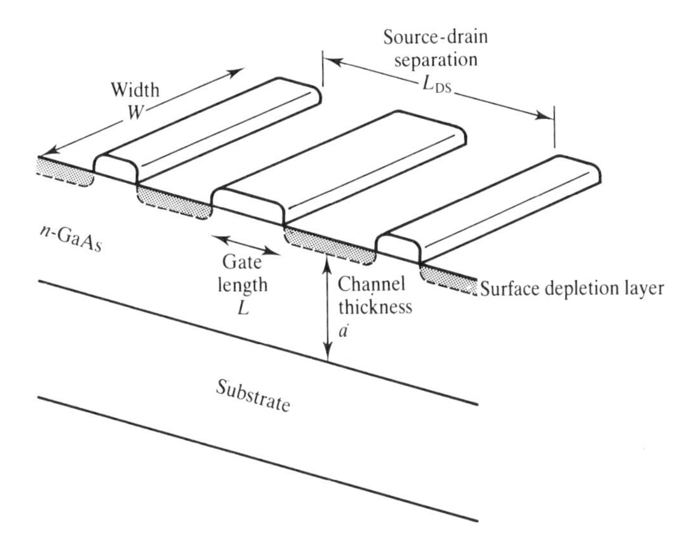

Figure 2.8 Dimensions in a MESFET

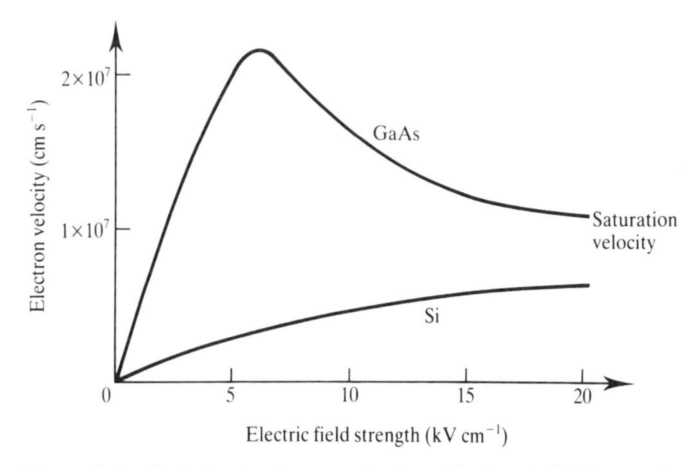

Figure 2.9 Variation in electron velocity with electric field strength in GaAs and silicon

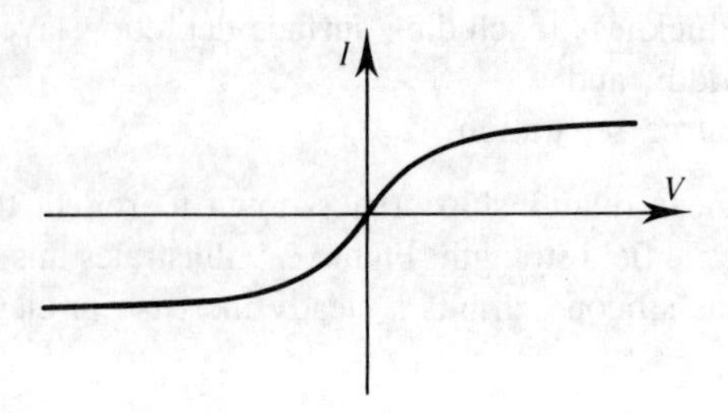

Figure 2.10 Current saturation in a GaAs resistor

strengths of the order of 20 kV cm^{-1}, GaAs has an electron velocity which peaks at relatively low fields and then falls (through a 'negative resistance' region) to a saturation velocity. The effect of this is to cause the current in a GaAs resistor to saturate at relatively low electric field strengths resulting in the $I-V$ characteristics shown in Figure 2.10. Saturation typically occurs at field strengths of approximately 5 kV cm^{-1}.

The importance of this current saturation can be seen if we consider a resistor structure fabricated with an ohmic separation of 10 μm. An electric field strength of 5 kV cm^{-1} will be attained with a potential difference of only 5 V, which is not an excessive level in integrated circuits. At greater field strengths, the resistor will not have a linear $I-V$ characteristic. The corollary of this is that if linearity is important, then the designer should consider the maximum field strength likely to be present across the resistor.

The parameter a in the equation above is the thickness of the conducting channel. It should be noted that this is not the same as the thickness of the doped region, since the energy states at the GaAs surface have an inherent charge which creates an associated depletion layer close to the surface. The surface depletion layer thickness is dependent upon the doping level in the substrate, but is typically of the order of 60 nm.

2.4.2 Gated devices (MESFETs)

The electrical characteristics of a contact between a metal and a semiconductor are usually classified as being either ohmic, where the current through the contact and the voltage across it obey a linear relationship, or rectifying, where the $I-V$ characteristics depend on the polarity of the applied potential (see Figure 2.11). In the rectifying case, the contact is known as a Schottky diode, and the current through the contact is given by the relationship:

$$I_G = A^*T^2LW \exp\left(-\frac{qV_{bi}}{kT}\right) \exp\left(\frac{qV_G}{nkT}\right)$$

where A^* (the effective Richardson constant) = 8.8 A cm K^{-1} @ 300 K;
T = junction temperature;
L = contact length;
W = contact width; and
n = ideality factor (usually in the range 1.1–1.2 for good diodes).

This equation is often approximated to:

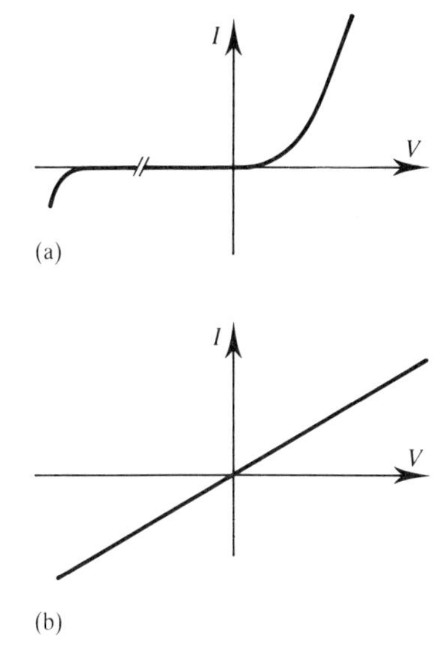

Figure 2.11 Electrical characteristics of (a) rectifying and (b) ohmic metal–semiconductor contacts

$$I_G \approx I_S \exp\left(\frac{qV_G}{kT}\right)$$

I_S is known as the saturation current and indicates the magnitude of the gate leakage current under reverse-bias conditions (V_G negative). V_{bi} is a constant known as the built-in voltage of the diode, and is dependent upon the materials used. V_G is the bias applied to the metal side of the Schottky diode, and for contacts to *n*-type semiconductors, V_G must be positive to forward bias the diode. As V_G approaches V_{bi} the resistance of the contact drops rapidly and appreciable conduction can occur. With negative (reverse) biasing, leakage currents are typically in the picoampere range (dependent upon the contact area). Schottky diodes on *n*-type GaAs generally have V_{bi} in the range 0.7–0.8 V. In comparison, Schottky diodes on silicon generally have V_{bi} in the range 0.4–0.6 V. The larger junction potential of GaAs Schottky contacts is advantageous in some digital ICs (e.g. DCFL, discussed further in Chapter 3) as V_{bi} effectively limits the logic swing and the noise margin in these circuits.

If a Schottky diode is positioned between two ohmic or $n+$ contacts, then by varying the bias on the Schottky contact it is possible to modulate the thickness of the diode depletion layer and thus the resistance of the channel between the ohmic contacts. This is the basis of operation of the MESFET structure described in Section 2.3; the ohmic contacts are the source and drain electrodes and the Schottky contact is the gate electrode. It is only in the region directly under the gate that resistance is modulated; this area is known as the active region of the MESFET. It should be noted that the structure is nominally symmetrical about the gate contact, although many variations have been proposed for non-symmetrical structures designed to improve various characteristics of the MESFET (e.g.

reference [12]). In the symmetrical structure, the choice of drain and source contact can be purely arbitrary. The source contact is generally defined as the ohmic contact having the most negative potential, and the gate and drain potentials are referenced to this.

It is possible to control the drain–source current (I_{DS}) by varying V_{DS} and V_{GS} in this structure. Two types of MESFET operation can be defined depending on whether or not the device is conducting at zero bias on the gate, which in turn depends upon the relative thicknesses of the doped region under the gate and the depletion layer thickness at zero bias. If the doped region is thinner than the zero bias depletion layer, the channel will be cut-off with zero gate bias and no drain–source current can flow. The device is thus normally-off and positive gate biases must be applied to reduce the depletion layer width so that a conducting channel can be formed. This enhancement-mode FET is often annotated as an E-MESFET (or an EFET). Conversely, if the doped region is thicker than the zero bias depletion layer, then the device is normally-on and negative gate biases must be applied to cut off the channel. This depletion-mode MESFET is often annotated as a D-MESFET (or a DFET). Symbols for the E-MESFET and D-MESFET are shown in Figure 2.12, and these are used throughout this book.

The measured I_{DS}–V_{DS} and I_{DS}–V_{GS} characteristics of a GaAs MESFET will be of the form shown in Figures 2.13 and 2.14. At small values of V_{DS}, the depletion layer under the gate can be considered to be a constant thickness dependent upon the gate bias (see Figure 2.15). The channel current obeys an approximately linear relationship with V_{GS} in this condition, so that the device behaves as a voltage-controlled resistor. A value, V_T can be defined at which the depletion layer and the channel are the same thickness, so that at gate biases more negative than this 'threshold voltage', the channel is cut off and no current can flow. The value of V_T thus indicates whether the device is depletion-mode or enhancement-mode. A positive value of V_T indicates that the channel is cut off until a positive gate bias (exceeding V_T) is applied; hence the device is an E-MESFET. Conversely, a negative V_T indicates that the device is a D-MESFET. Circuits fabricated from MESFETs with a large negative V_T will generally be faster than those with a smaller (more positive) V_T due to the lower resistance associated with the thicker channel, and hence the faster charging and discharging of the circuit capacitances.

At higher values of V_{DS} the gate–channel voltage will vary, dependent upon the position along the channel, with the gate–channel potential difference at the drain end

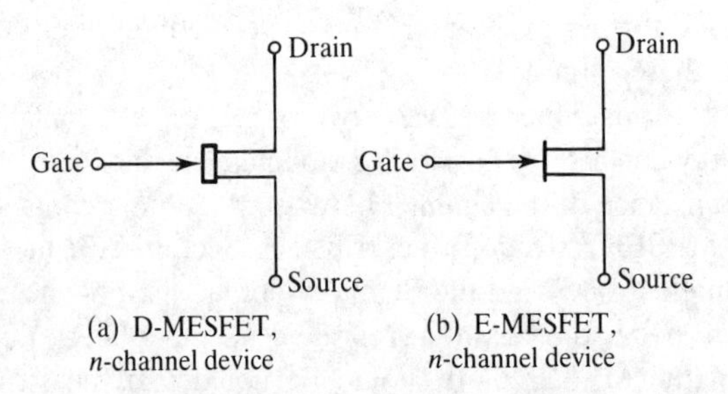

Figure 2.12 MESFET symbols: (a) D-MESFET, *n*-channel device; (b) E-MESFET, *n*-channel device

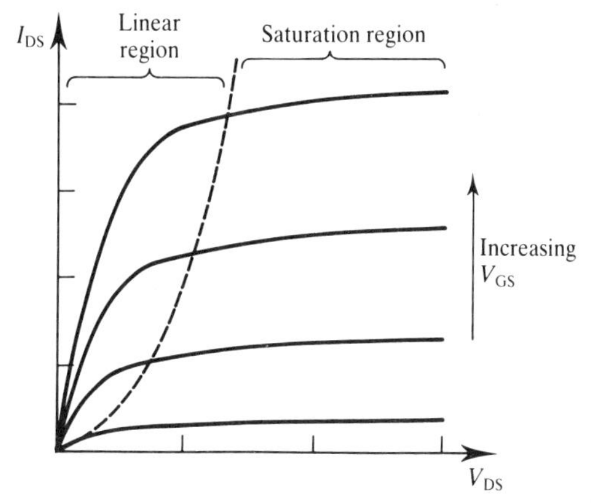

Figure 2.13 I_{DS}–V_{DS} characteristics of a GaAs MESFET

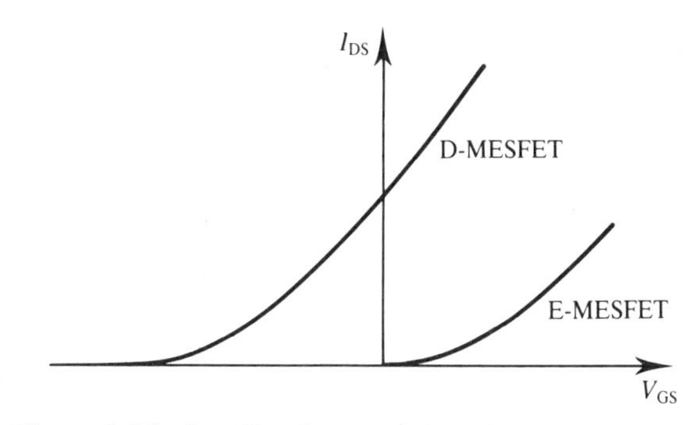

Figure 2.14 I_{DS}–V_{GS} characteristics of a GaAs MESFET

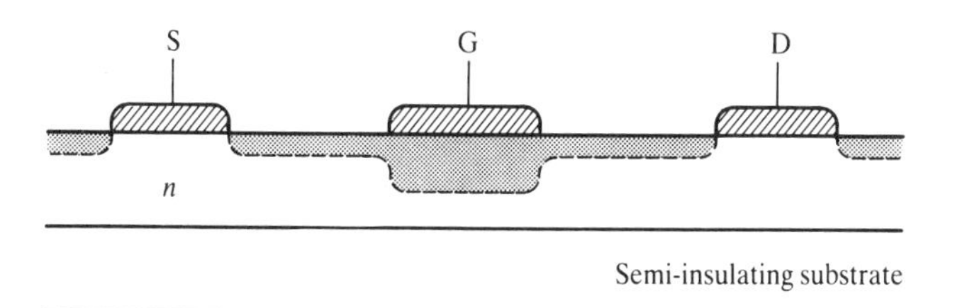

Figure 2.15 The gate depletion layer in a MESFET in the 'linear' region: S, source; G, gate; D, drain

becoming increasingly more negative (or less positive) as V_{DS} is increased. The depletion layer width will thus also vary with position, as shown in Figure 2.16. As V_{DS} increases the depletion layer width at the drain end will increase until it is the same width as the channel. In this situation, further increases in V_{DS} only lead to small increases in I_{DS}; hence this is referred to as the 'saturation' region. The channel current does not saturate to a constant value in this region (as might be expected from the name) since a high electric

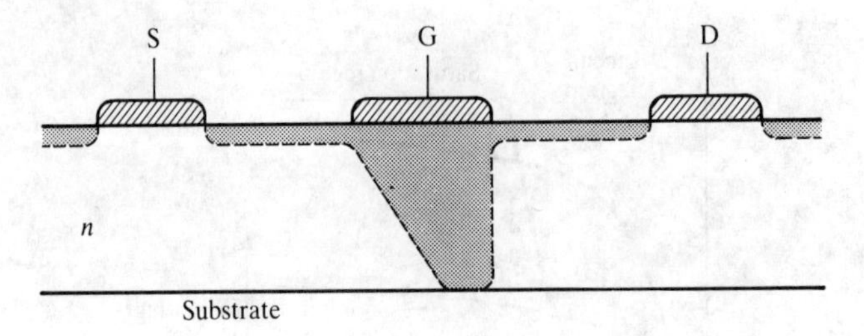

Figure 2.16 The gate depletion region in a MESFET in the 'saturation' region: S, source; G, gate; D, drain

field exists in the pinched-off area of the channel, and this (combined with leakage through the substrate) causes some slight dependency of I_{DS} upon V_{DS} in the saturation region.

In the saturation region I_{DS} is approximately proportional to the square of $(V_{GS} - V_T)$. The threshold voltage is thus commonly measured by extrapolating $(I_D)^{1/2}$ vs V_{GS} onto the V_{GS}-axis and defining the intercept as V_T, although some workers have used alternative definitions (such as the value of V_{GS} giving I_{DS} equal to 1 percent of I_{DSS} which is the value of I_{DS} at zero gate bias). Publications describing GaAs MESFET ICs often quote the threshold voltage as a means of comparison; it should be noted, however, that some (mainly older) publications may quote the process 'pinch-off' voltage instead. Strictly speaking the pinch-off voltage is the drain–source voltage which extends the depletion layer across the channel at the drain-end of the gate (not the gate–source voltage at which the whole channel is cut off as is used for the threshold voltage definition). The pinch-off voltage defines the transition from the linear region to saturation and is given by $V_P = (V_{GS} - V_T)$. For the case $V_{GS} = 0$ V then $V_P = (-V_T)$; however, since V_P is a function of V_{GS}, it is more useful to use V_T to compare GaAs MESFET devices. Most modern references now follow this notation.

To a good approximation the I–V characteristics of a GaAs MESFET can be described by the following equations:

$$I_{DS} = \beta(2[V_{GS} - V_T] - V_{DS})V_{DS}(1 + \lambda V_{DS})$$

in the linear region, and

$$I_{DS} = \beta(V_{GS} - V_T)^2(1 + \lambda V_{DS})$$

in saturation. In these equations β and λ are constants dependent upon the MESFET properties. β and thus I_{DS} are proportional to $(\mu_n W/L)$; it is this dependence on the electron mobility, μ_n, which gives GaAs ICs their relatively higher speed compared to silicon ICs. Since β is also inversely proportional to the gate length L, GaAs devices are usually fabricated with as short a gate length as the technology will allow; devices with long gate lengths would only operate as fast as shorter gate length silicon devices which could be fabricated at lower cost and higher yield. Submicron gate lengths are now commonly used.

The constant λ is a measure of the output conductance (g_o) of the MESFET. High values of λ correspond to high values of g_o, and relatively poor saturation. The measured

$I-V$ characteristic is not flat in saturation due to the combined effects of various leakage currents and the high electric field present at the drain end of the channel, and λ is an indication of the magnitude of these leakage currents.

From the sensitivity of I_{DS} to V_{GS} and V_{DS}, it is possible to define two conductances given by:

(a) the drain (or output) conductance

$$g_o = \frac{\partial I_{DS}}{\partial V_{DS}}$$

at constant V_{GS} and

(b) the transconductance

$$g_m = \frac{\partial I_{DS}}{\partial V_{GS}}$$

at constant V_{DS}.

If we use the simple JFET equations to model the MESFET characteristics, then we can calculate the transconductance g_m and the output conductance g_o in the saturation region as:

$$g_m = \frac{2 I_{DS}}{(V_{GS} - V_T)}$$

and

$$g_o = \frac{\lambda}{(1 + \lambda V_{DS})} I_{DS}$$

It can be seen that the value of transconductance is strongly dependent upon V_{GS}. Commonly measurements of g_m will be made at zero gate bias for D-MESFETs and +0.7 V gate bias for E-MESFETs.

The transconductance and output conductance parameters are important as they define the dependence of the MESFET current to small a.c. signals. These conductances are used in the a.c. small-signal hybrid-pi model of the MESFET which is discussed further in Chapter 4. The conductance ratio g_m/g_o is a common factor determining the performance of many analog circuits; typically, this ratio is in the range 15–25.

2.4.3 MESFET capacitance

The speed of a GaAs MESFET circuit is dependent upon the speed with which the charge within the MESFETs can be modulated. This charge must be modulated through various parasitic resistances associated with the device structure. The charge is primarily stored in capacitance associated with the depletion layer underneath the gate. If the assumption is made that the channel has a uniform doping profile then the depletion layer width is given by:

$$x_d = \left(\frac{2\epsilon_r \epsilon_o V_{bi}}{qN_D} \right)^{1/2}$$

The gate capacitance can then be evaluated by treating the gate-depletion layer–channel structure as a parallel plate capacitor of separation x_d. At zero drain bias, the gate capacitance can be considered as being equally shared between the source and drain contacts in symmetrical structures. The total gate capacitance is typically of the order of 1 fF μm^2 at a channel doping density of 10^{17} cm^{-3} and zero gate–channel bias; this capacitance is inversely proportional to the square of the gate reverse-bias voltage in a uniformly doped structure. In practice, the channel doping profile will not be uniform (particularly if ion implantation is used for processing, see Chapter 5), but the above value of gate capacitance is still a good approximation.

2.4.4 Source and drain parasitic resistance

In any MESFET device, it is only the channel immediately underneath the gate electrode which is modulated by the gate bias, and strictly speaking it is only this 'intrinsic' portion of the MESFET which obeys the I–V relationships given above. The regions either side of the gate form parasitic resistances connecting the intrinsic MESFET to the outside world. Any current flowing through these resistances will produce a potential drop so that the values of V_{GS} and V_{DS} present in the intrinsic device will be less than those actually applied at the contacts. The value of I_{DS} produced by an applied V_{GS} and V_{DS} will thus be degraded by the parasitic resistance, and the corresponding transconductance will likewise be degraded. To a first approximation the extrinsic transconductance is given by:

$$g_m = \frac{g_{mo}}{1 + r_s g_{mo}}$$

where g_{mo} is the intrinsic value and r_s is the parasitic resistance on the source side of the gate.

Any degradation in I_{DS} and g_m will have a corresponding effect on the speed [13] of the IC, so one of the main aims of a GaAs processing engineer is to minimize the parasitic resistances present in the MESFETs. Methods for achieving this chiefly entail using either a recessed gate structure or a self-alignment technology. These techniques are described in some detail in Chapter 5.

As an example of the impact of parasitic source and drain resistance on a MESFET, consider the structure shown in Figure 2.17. The open-channel resistance under the gate is given by the relationship:

$$r_{channel} = \frac{L}{qN_D \mu_n a W} = 8.3\ \Omega$$

The total source parasitic resistance is given by:

$$r_s = r_c + r_{s1}$$

where r_{s1} can be considered approximately equal to $r_{channel}$ for this device (since the

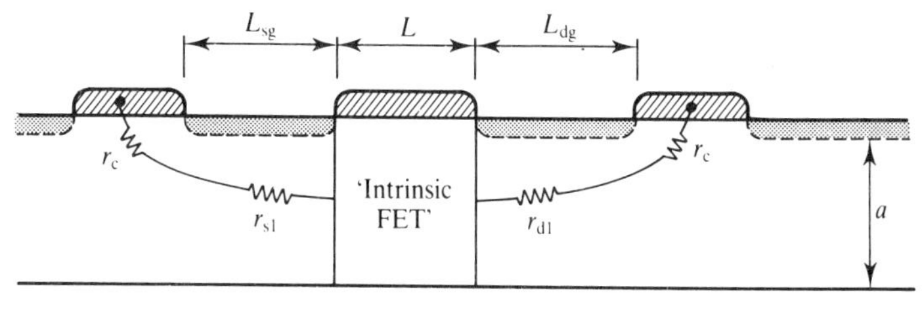

Figure 2.17 Parasitic resistances in a GaAs MESFET (N_D = 10^{17} cm^{-3}, μ_n = 3000 cm^2 V^{-1} s^{-1}, W = 100 μm, a = 0.25 μm, $L_{sg} = L_{dg} = L$ = 1 μm)

ohmic-gate separation is the same as the gate length) if we neglect the effect of the surface depletion layer. The contact resistance r_c is given by:

$$r_c = \frac{1}{W}\left(\frac{\rho_c}{q\mu_n N_D a}\right)^{1/2}$$

(ρ_c is the specific contact resistance which is approximately 10^{-6} ohm cm^2 for the AuGe ohmic contacts commonly used).

Calculating r_s gives:

$$r_{s1} = 8.3\ \Omega$$
$$r_c = 2.9\ \Omega$$

and thus r_s = 11.2 Ω. Since the structure is symmetrical, the total parasitic drain resistance r_d = 11.2 Ω also. The total channel resistance of this structure is thus 30.7 Ω, of which it is only possible to modulate an 8.3 Ω section immediately underneath the gate. It should be evident that the device characteristics will be poor in comparison to an ideal device having no parasitic resistance.

2.4.5 Gate parasitic resistance

Although there will be little direct current flowing through the gate contact under reverse-bias conditions, a current will flow in response to any gate signal changes as the gate capacitance is charged or discharged through the gate resistance. The finite gate resistance means that changes in the potential at the far end of the gate will be delayed with respect to changes in the potential of the near (signal) end, and this will degrade the output rise and fall times since the channel thickness will not be modulated as a single 'lumped' component. The parasitic gate resistance also contributes to the noise produced by the MESFET.

Methods of reducing the parasitic gate resistance involve either using multiple gate fingers or contact points, as shown in Figure 2.18, or using special processing to give the gate a 'T' or 'mushroom' cross-section. This structure has a narrow bottom which allows a short gate length to be maintained, while a wider top section reduces the overall gate

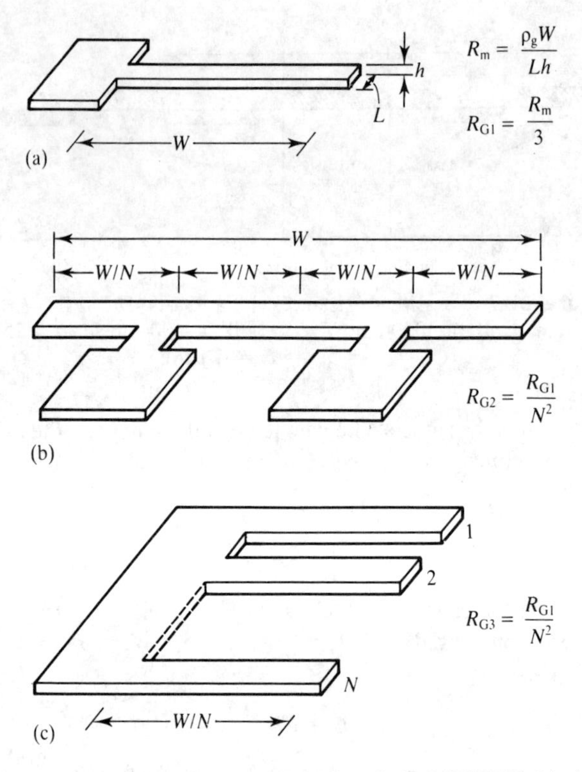

Figure 2.18 Parasitic gate resistance in a GaAs MESFET: (a) single pad; (b) multiple pads; and (c) multiple fingers

resistance. This gate cross-section is used in the SAINT self-alignment process which is described in Chapter 5.

2.4.6 Output conductance frequency dependence

If the output conductance g_o is measured at different frequencies, it will be seen to exhibit an appreciable dependence on frequency at low frequencies (decreasing with decreasing frequency) and to be constant only above frequencies of the order of one kilohertz [14] (see Figure 2.19). It is thought that the variation can be attributed to charge exchange with the surface states in GaAs MESFETs [15]. This effect has considerable importance if device parameters are measured at low frequency (e.g. by using a curve tracer or semiconductor parameter analyser) while circuit analysis is required to determine the circuit performance at high frequencies. It is important that MESFET parameter measurements are made at a frequency relevant to the analysis to be performed. The variation in output conductance can be modelled either through the use of different values of λ appropriate to the different frequencies concerned, or by adding an RC network in parallel with the MESFET drain–source channel, as shown in Figure 2.20.

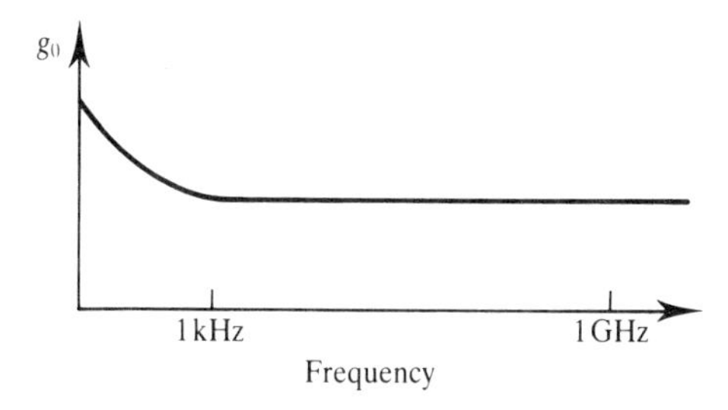

Figure 2.19 Variation in output conductance with frequency in a GaAs MESFET

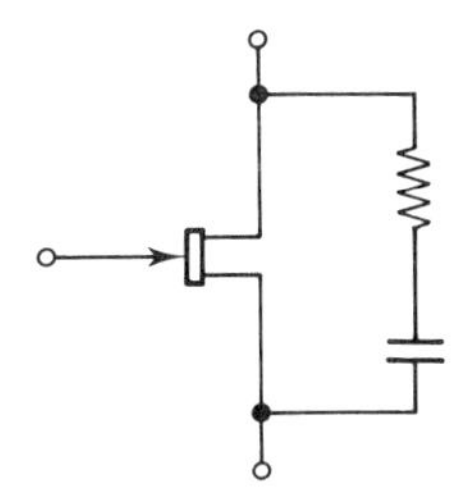

Figure 2.20 MESFET model amended to include the frequency dependence of the output conductance

2.4.7 Effect of scaling device dimensions

It may be remembered that the saturated drain current in a GaAs MESFET can be approximated by:

$$I_{DS} = \beta(V_{GS} - V_T)^2(1 + \lambda V_{DS})$$

The parameter β in this equation is proportional to $(W\mu_n/L_g)$, where W and L_g are the MESFET gate width and length, respectively, and μ_n is the electron mobility.

As the gate lengths increase, so β and thus I_{DS} decrease, while the gate capacitance increases in proportion to L_g. In a digital circuit the propagation delay is dependent on both the capacitance in the circuit and the current available to charge or discharge this capacitance, so we can say that:

$$\text{Propagation delay} \propto \frac{C}{I} \propto (L_g)^2$$

The power dissipation of each logic gate is proportional to I_{DS}, and thus:

$$\text{Power} \propto \frac{1}{L_g}$$

and the power-delay product is proportional to L_g.

Reducing the gate length dimension thus reduces the power-delay product, as shown

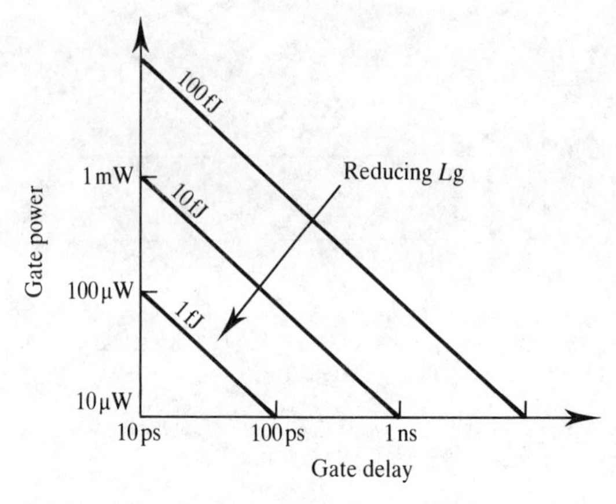

Figure 2.21 Reduction in the power-delay product through reduced gate length

in Figure 2.21, and this is the principal motive behind the drive towards smaller and smaller device features in all integrated circuit technologies. A reduction in the power-delay product can be obtained by reducing the gate length until the determination of the gate propagation delay is dominated by stray capacitance, e.g. from interconnecting tracks. At this point, all the other feature sizes (e.g. ohmic contact areas, track widths, etc.) must be reduced before any further reduction in the power-delay product can be obtained.

This dependence of the power-delay product on gate length is the reason why GaAs gate lengths have moved to submicron dimensions. ICs using longer gate lengths would only have the same performance as a short gate length silicon IC, which could be fabricated at less cost and higher yields. As was discussed in Chapter 1, the cost and yield factors effectively limit the use of GaAs ICs to state-of-the-art applications using state-of-the-art dimensions to attain a level of performance that cannot be achieved by other means.

2.4.8 Short-channel effects

If very high speed circuits are required, it is usual to reduce the MESFET gate length to below one micron. At these submicron gate lengths the high electric fields present in the channel cause the electron velocity to overshoot the peak velocity, resulting in a lower effective resistivity. This speed increase is often accompanied by a change in the characteristics known as the 'short-channel effect'. Predominantly this change appears as a negative shift in the threshold voltage of the device. The origin of this threshold voltage shift is thought to be related to the presence of leakage currents in the substrate which become significant under the influence of the very high electric fields in a short gate length device [16]. Some success at suppressing the effect has been achieved by the addition of a buried *p*-layer beneath the conducting channel [17].

2.5 Analysis of GaAs circuits

Detailed analysis of the behaviour of a GaAs IC requires the solution of the $I-V$ equations for each of the components in order to determine the operating conditions and the response to any external signals, taking into account the effect of the capacitances within the devices and circuit which will limit the speed with which signals can change. Theoretically, this analysis could be carried out manually; however, for any reasonably useful circuit the number of equations would make this analysis extremely difficult and time-consuming. One solution to this problem is to use simplified models (such as the 'hybrid-pi' model discussed in Section 4.1) for each device. This approach is recommended to allow the designer to gain 'a feel' for how the circuit operates and for which parameters are dominant in determining the overall circuit characteristics (see Chapter 4). If a detailed analysis is required, however, this method will not give sufficient accuracy. The alternative is to use a computer program to analyse the circuit. Many commercial programs exist to carry out this task, the most well-known and widespread throughout industry and academia being SPICE [18].

The standard version of SPICE does not include a specific model for GaAs MESFETs; however, a silicon JFET model is included which uses the same equations described in Section 2.4.2. If appropriate parameters are used in this model then the predicted current can agree well with the measured current in the saturation and linear regions, although there is some error in the area of the 'knee' at the transition between these regions [19]. This error takes the form of a transition to the saturation region at a lower voltage than predicted by the simple JFET model; it is sometimes referred to as 'early saturation'. An analog circuit will commonly be operated with all its components in the saturation region, so the JFET equations allow a good indication of analog circuit performance provided the variation in output conductance with frequency is allowed for by using a value of λ appropriate to the signal frequency concerned. Analysis of a digital circuit would give reasonably accurate predictions in the logic 'high' and 'low' conditions, but would underestimate the current available in the transition between these states and thus underestimate the circuit speed.

Improved accuracy can be obtained by modifying the equations, and many commercial versions of SPICE (e.g. PSPICE and HSPICE) include specific models for GaAs MESFETs. The most commonly used modification is that proposed by Curtice [20], which uses an equation of the form shown:

$$I_{DS} = \beta(V_{GS} + V_T)^2(1 + \lambda V_{DS})\tanh(\alpha V_{DS})$$

Additional details of the equation modifications required to yield further accuracy improvements are beyond the scope of this book, but can be found in the literature [21]. It should be emphasized that it is not always necessary to seek extreme accuracy, nor is it always meaningful when one considers that real devices will exhibit a range of characteristics across a wafer and will not all be identical as assumed by the simulation program. Very useful results can be achieved with the simple JFET model provided the circuit designer remembers that the accuracy is limited.

Table 2.1 lists parameters for the SPICE JFET model for 1 μm gate length D-MESFETs produced by various GaAs IC manufacturers. Representative parameters for a typical 1 μm gate length E-MESFET are shown in Table 2.2. All the parameters shown in these tables have been scaled appropriately for a 1 μm gate width. Where two values of λ are given, this reflects the dependence of output conductance upon frequency; the higher value should be used for frequencies above 1 kHz, the lower values for d.c. analysis. Schottky diodes can be modelled using these SPICE parameters through the use of a MESFET with its source and drain connected together, as shown in Figure 2.22.

Table 2.1 SPICE JFET parameters for GaAs D-MESFET modelling (L_G = 1 μm, W_G = 1 μm)

Parameter	Manufacturer	
	STC Technology Ltd [22]	Gigabit Logic Inc. [23]
VTO (V)	−1.0	−0.9
β (10^{-6} S V^{-1})	67	148
λ (V^{-1})	0.3 (a.c.)	0.023 (d.c.) 0.13 (a.c.)
R_S (Ω)	2920	1055
R_D (Ω)	2920	1055
IS (10^{-15} A)	0.075	20
PB (V)	0.79	0.85
FC	0.9	—
CGS (10^{-15} F)	0.39	1.3
CGD (10^{-15} F)	0.39	0.62

Table 2.2 SPICE JFET parameters for GaAs E-MESFET modelling (L_G = 1 μm, W_G = 1 μm)

Parameter	Value
VTO (V)	0.2
β (10^{-6} S V^{-1})	130
λ (V^{-1})	0.3
R_S (Ω)	4520
R_D (Ω)	4520
IS (10^{-15} A)	0.075
PB (V)	0.79
FC	0.9
CGS (10^{-15} F)	0.39
CGD (10^{-15} F)	0.39

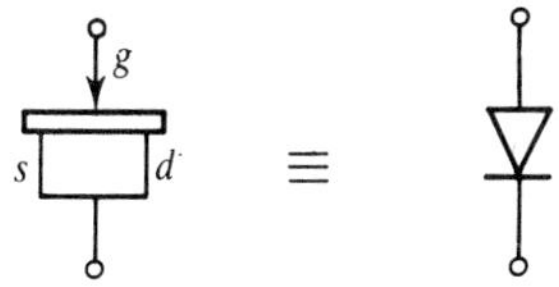

Figure 2.22 Schottky diode model based upon a GaAs MESFET

3 □ Digital design

3.1 Inverters

The most basic function required in any logic circuit is inversion. This function is commonly implemented by using the input signal to control the current which flows through a load device connected to a positive supply rail (see Figure 3.1). When the current is 'on', a potential difference will be developed across the load and thus the output voltage will be low (logic '0'). When the current is 'off', no potential difference will be developed, and the output voltage will be high (logic '1'). To achieve inversion the current must be controlled by some (switch) device which turns the current 'on' when the input voltage is high (logic '1'), and turns it 'off' when the input is low (logic '0').

The switch device must be able to control the current in accord with the input voltage level, in GaAs IC technology this function can be achieved by a MESFET and a choice can be made between using an E-MESFET or a D-MESFET for this device. The load must develop a potential difference when a current passes through it, and either a resistor or a MESFET will fill this criterion. Independent of which devices are used, however, the inverter output voltage will swing between a logic high (V_{HI}) equal to the positive supply voltage V_{DD}, and a logic low (V_{LO}) determined by the ratio of the load resistance (R_{LOAD}) and the on-resistance of the switch (R_{ON}):

$$V_{LO} = V_{DD}\frac{R_{ON}}{R_{ON} + R_{LOAD}}$$

$$= \frac{V_{DD}}{1 + R_{LOAD}/R_{ON}}$$

The level of V_{LO} must be low enough to turn off the next logic gate in the circuit. It is thus very important that the ratio R_{LOAD}/R_{ON} is kept high enough to ensure that the next stage is turned off properly, even when the IC parameters are subject to the inevitable random variations which occur in processing. Since R_{ON} is determined by the properties

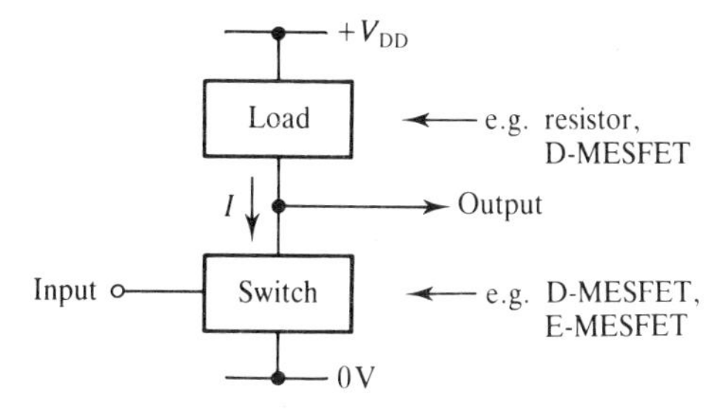

Figure 3.1 The ratioed-logic inverter structure

of a MESFET, the load device is usually also chosen to be a MESFET rather than a resistor, as the value of V_{LO} is then determined by the parameters of two MESFETs rather than a resistor and a MESFET. Better tracking between the values of R_{LOAD} and R_{ON} is then achieved. A further advantage is given by using a MESFET in that the depletion layer under the gate metal in a MESFET is thicker than under the bare GaAs surface (see Chapter 2), so that a MESFET has a higher resistivity than a resistor of similar dimensions. The physical size of a MESFET used for R_{LOAD} will be smaller than a resistor having the same resistance, and thus a higher packing density (gates per unit area) can be achieved using MESFET loads. This is important if VLSI levels of integration are to be achieved.

When connecting the load MESFET into the inverter circuit, the question arises as to what to do with the third (gate) terminal. Strictly speaking it is only necessary to pass current between the drain and source contacts to develop a voltage drop which can be used for the output level. However, if we leave the gate contact floating, we effectively have a resistor device again and we lose the advantages given by a MESFET load. If we connect the gate to the positive rail (V_{DD}), as shown in Figure 3.2(a), the gate–source diode becomes forward-biased and we then effectively have a diode for a load instead of a MESFET. The logic low available from this inverter will be limited to approximately 0.7 V (the forward voltage drop of the Schottky diode) below V_{DD}. If we try to pull V_{LO} below this level, then the gate–source diode in the load becomes forward-biased and effectively clamps V_{LO}. This clamped logic low will generally be too high to turn off the next logic gate correctly.

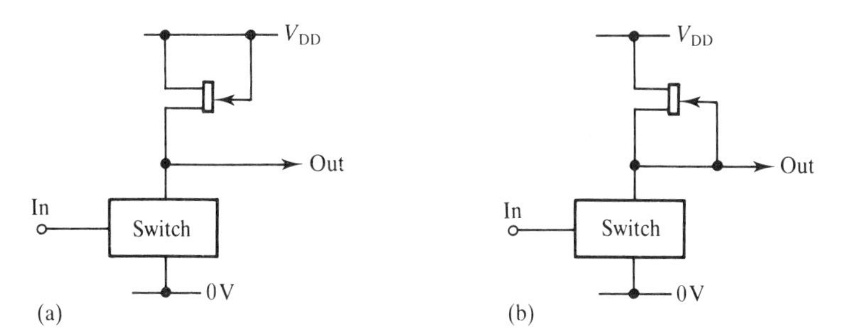

Figure 3.2 Possible load-FET connections in an inverter: (a) gate–drain connection; (b) gate–source connection

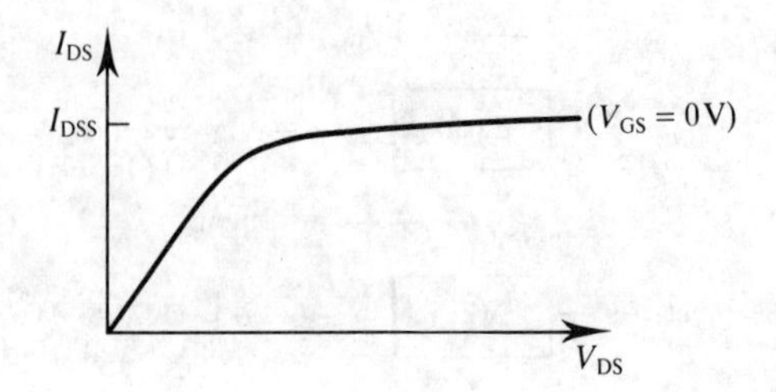

Figure 3.3 I–V characteristics of a load-FET

The only practical alternative is to connect the gate to the source contact (see Figure 3.2(b)). With this connection, the MESFET operates under a fixed zero gate–source voltage ($V_{GS} = 0$) condition, and the I–V characteristic is of the form shown in Figure 3.3. It should be obvious from these I–V characteristics that the MESFET used for the load must be a depletion-mode device; an enhancement-mode device will have a zero drain–source current (I_{DS}) when $V_{GS} = 0$, and this zero current will prevent the inverter from operating.

The use of a D-MESFET rather than a resistor for the load also gives a speed advantage which can be seen by examining the I–V characteristics of the switch device with load-lines superimposed (Figure 3.4). The load-line shows how the voltages across the switch and the load are related by the common current flowing through them. The load-line is formed by reversing and scaling the I–V characteristic of the load onto the switch I–V characteristic so as to fit the boundary conditions:

(a) when the current is zero, the load potential drop is zero and the switch potential drop is V_{DD}, and
(b) when the current is at a maximum (determined by V_{DD}/R_{LOAD}) the load potential drop is V_{DD} and the switch potential drop is zero.

At values of current between these limits, the voltages across the load and the switch must sum to V_{DD}, and each can be read off the graph. Since the current is determined

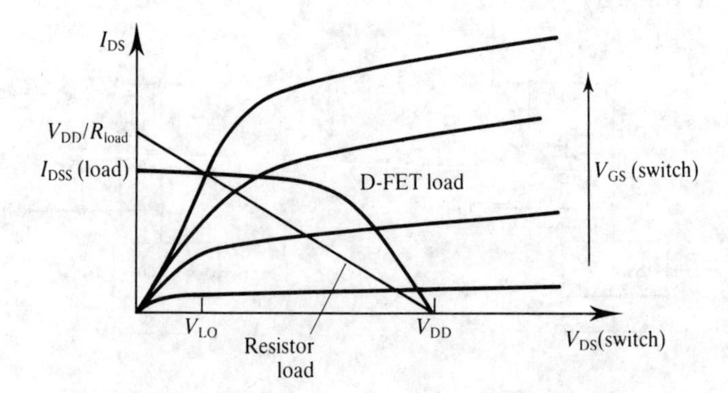

Figure 3.4 Load-line characteristics for an inverter

by the input voltage, the load-line can thus be used to determine the dependency of the output voltage upon the input voltage (this is the transfer characteristic of the logic gate which is referred to further in Section 3.7). The load-line also shows how much current is available for charging any capacitance attached to the output; even if the switch device is capable of turning-off instantaneously, the output voltage will not rise at the same rate but will be limited by the rate at which the output capacitance can be charged through the resistance of the load device. The load current thus determines the circuit speed. Unfortunately, fast circuits cannot be constructed simply by using small resistor values since this will affect the logic low voltage (and may prevent the circuit working correctly). Maximum speed has to be achieved with a load which gives the required logic low voltage with the maximum area under the load-line. Figure 3.4 shows load-lines for a D-MESFET and a resistor which give the same logic low value. It can be seen that the non-linear characteristic of the MESFET load gives a significant advantage in terms of the current available. (NB. If very small resistor lengths are used so that current saturation occurs due to the high electric fields present in these devices, then a similar speed improvement can be obtained, although these very small resistor dimensions are difficult to fabricate [24, 25].)

Up to this point no distinction has been made between the use of an E-MESFET or a D-MESFET for the switch device; the above discussion applies equally to either option. The inverter will operate correctly provided that the input signal obeys the following two constraints:

(a) In order to turn the current off, the input voltage must be taken below the switch threshold voltage.
(b) To turn the current on, the input voltage must be taken above the threshold voltage to a level sufficient to produce an on-resistance and thus a logic low output able to turn off the next logic gate.

In order to meet these constraints the choice of E-MESFET or D-MESFET for the switch device is responsible for determining the final inverter circuit and the logic operating parameters such as the logic voltage levels, propagation delays and the power dissipation. The following sections distinguish between the options, and describe the relevant design details for the resulting logic gate families.

3.2 Direct-coupled FET logic (DCFL) [26]

The simplest form of the GaAs inverter uses an E-MESFET for the switch device, as shown in Figure 3.5. The output logic high voltage will move towards V_{DD} when the switch is off, but be clamped by the forward gate–source diode drop at the input of the next switch device to give an approximate $V_{HI} = 0.7$ V. The logic low voltage will be determined by:

$$V_{LO} = \frac{V_{DD}}{1 + R_{LOAD}/R_{ON}}$$

as discussed previously, where R_{ON} is the on-resistance of the E-MESFET having a gate

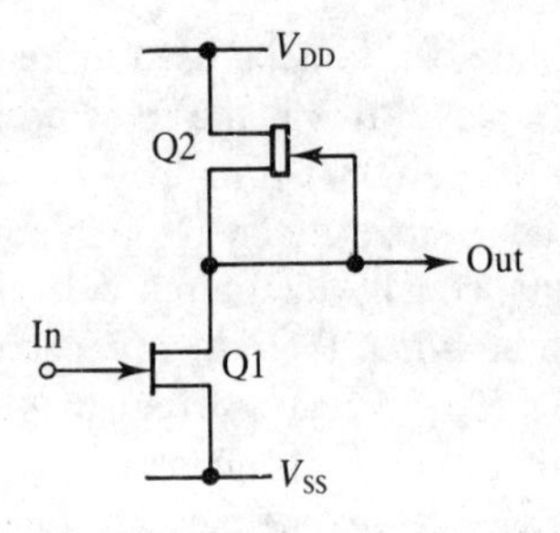

Figure 3.5 A direct-coupled FET logic (DCFL) inverter (typical parameters: $V_{DD} = 1.7$ V, $V_{SS} = 0$ V, $W_{Q2} = 5\ \mu$m, W_{Q1} dependent upon threshold voltages — see text)

voltage of V_{HI}. The key to correct operation of the logic circuit is that V_{LO} must fall below the threshold voltage of the E-MESFET switch. Design of the inverter involves scaling the device widths in line with the threshold voltages for the load and the switch, and the supply voltage so as to produce a good logic low.

To calculate the appropriate widths of the switch and load, the approximate equations for GaAs MESFETs given in Chapter 2 can be used. Reiterating these we have:

$$I_{DS} = \beta V_{DS}[2(V_{GS} - V_T) - V_{DS}](1 + \lambda V_{DS})$$

in the linear region ($V_{DS} < (V_{GS} - V_T)$), and

$$I_{DS} = \beta(V_{GS} - V_T)^2(1 + V_{DS})$$

in the saturation region ($V_{DS} > (V_{GS} - V_T)$). When the input voltage of the inverter of Figure 3.5 is V_{HI}, and the output voltage is V_{LO}, then the E-MESFET switch will be in its linear region of operation, while the D-MESFET load will be in its saturation region.

Since we must have current-continuity in this circuit, we can equate the currents flowing in the E-MESFET and the D-MESFET to get:

$$\begin{aligned} I_{DS} &= \beta_E V_{LO}[2(V_{HI} - V_{TE}) - V_{LO}](1 + \lambda_E V_{LO}) W_E \\ &= \beta_D(-V_{TD})^2(1 + \lambda_D[V_{DD} - V_{LO}]) W_D \end{aligned}$$

In deriving this equation, V_{DS} and V_{GS} have been replaced by V_{HI}, V_{LO} or zero volts as appropriate, and the drain currents have been scaled by factor W, representing the width of the MESFET. The subscripts E and D refer to the E-MESFET switch and the D-MESFET load, respectively.

The output low must be below the threshold voltage of the switch device to be able to turn off the next stage. The supply rail V_{DD} is often derived from those required to power silicon emitter-coupled logic circuits (ECL) since it is assumed that most systems will have a mixture of GaAs and silicon ICs, and it is desirable if the total number of supply rails can be minimized; V_{DD} is thus set to 1.7 V above V_{SS} (see Section 3.11 for further explanation). The E-MESFET threshold voltage is usually chosen to be approximately 10 percent of the supply voltage, i.e. around 0.2 V. V_{LO} must fall below this

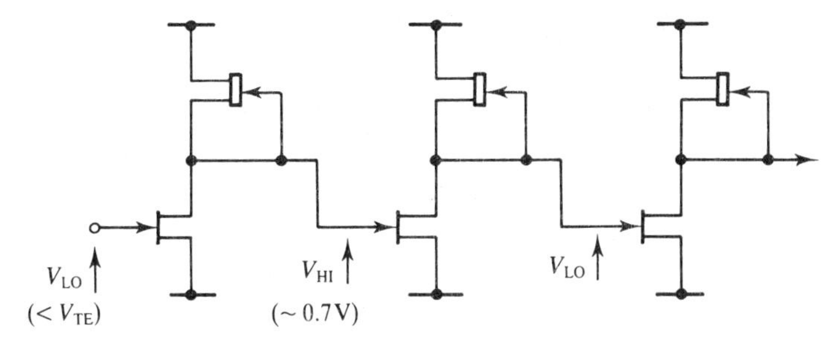

Figure 3.6 Logic high and low voltages in DCFL

threshold voltage to be able to turn the E-MESFET off, typically it is required to be approximately 0.1 V. The output high V_{HI} is limited by the forward voltage drop of the gate–source diode at the input of the next stage (see Figure 3.6), typically this is approximately 0.7 V.

Using these values, we can write:

$$\frac{W_E}{W_D} = \frac{\beta_D V_{TD}^2(1 + \lambda_D 1.6)}{\beta_E 0.1[2(0.7 - V_{TE}) - 0.1](1 + \lambda_E 0.1)}$$

If we assume that $\beta_D = \beta_E$ and that $\lambda_D = \lambda_E = 0.15\ V^{-1}$ (these are approximations based on typical values — see Chapter 2), then

$$\frac{W_E}{W_D} = \frac{(V_{TD}^2)1.24}{(0.13 - 0.2V_{TE})1.015}$$

Example values of the MESFET width ratio calculated by this formula are given in Table 3.1. These figures show that for the maximum number of gates per unit area (i.e. maximum packing density), the threshold voltage of the load should be close to -0.3 V. Large (negative) values of V_T will necessitate the use of wide switch devices and hence the consumption of large amounts of chip area. Although these rough calculations indicate the approximate width ratio, detailed calculations should also be carried out (using the SPICE simulator, for example) so as to maximize the gate noise margin (this is discussed in more detail in Section 3.7).

When correctly designed, the output levels of this logic gate allow direct coupling to the inputs of others, and thus logic designed using this configuration is known as directly-coupled FET logic or DCFL.

Table 3.1 Typical FET width ratios for DCFL: V_T (switch) $= +0.2$ V, $V_{DD} = 1.7$ V

V_T (load)	-1.0	-0.8	-0.4	-0.3
W_E/W_D	13.6	8.7	2.2	1.2

3.3 Buffered FET logic (BFL) [27]

Although conceptually simple, DCFL circuits are difficult to make due to the need to integrate both enhancement and depletion-mode MESFETs. Historically, the first digital GaAs ICs to be fabricated used D-MESFET devices only. The use of a D-MESFET switch means that the input voltage must swing negative (below V_T) to turn the inverter current off; however, the output of the basic inverter structure (shown in Figure 3.1) cannot fall below zero volts. In order to be able to drive a subsequent logic gate, the output must be level-shifted negatively, and this is achieved by adding a level-shift stage, as shown in Figure 3.7.

In this level-shift, stage Q3 and Q4 are equal-sized MESFETs, and in order to achieve equal currents through both devices, V_{GS} of Q3 must thus be approximately equal to V_{GS} of Q4. (Approximately, because a different drain–source voltage may exist across each device.) V_{GS} of Q3 must thus be approximately zero, and the source voltage of Q3 will follow any changes in the gate voltage. (Hence this stage is also known as a source–follower, similar to the emitter–follower circuit commonly used in silicon bipolar circuits.) If the source voltage of Q3 follows changes in its gate voltage, then so will the voltage at the bottom of the diode chain, although at a more negative d.c. level determined by the number of diodes in the chain. The output voltage is thus level-shifted down from the level at the output of the inverter stage.

By making Q3 and Q4 larger than the inverter devices Q1 and Q2, a greater capacity for driving loads of low impedance or high capacitance is conferred. The level-shift stage thus buffers the inverter from the load as well as level-shifting the output voltage. The complete design of this inverter is thus commonly referred to as buffered FET logic, or BFL.

The design of a BFL gate in detail demands a knowledge of the characteristics of the

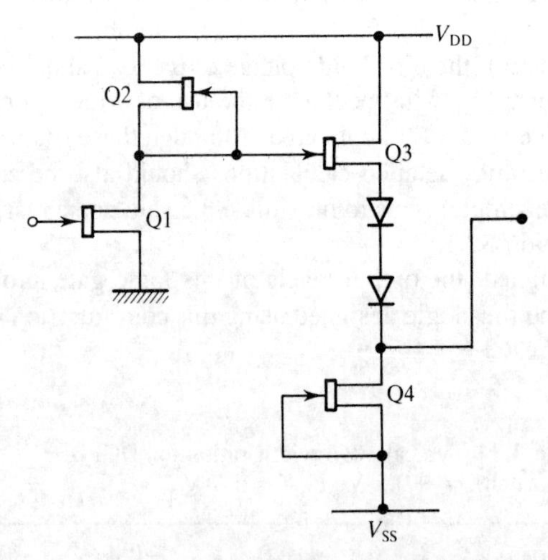

Figure 3.7 A buffered FET logic (BFL) inverter (typical parameters: $V_{DD} = 3$ V, $V_{SS} = -2$ V, $W_{Q1} = 10\ \mu$m, $W_{Q2} = 5\ \mu$m, $W_{Q3} = 20\ \mu$m, $W_{Q4} = 20\ \mu$m)

D-MESFETs and diodes being used so that the noise margin (see Section 3.7) can be optimized. However, a rough-and-ready design can be made by a series of very simple calculations; this rough design can be used as the starting point for more detailed analysis. The stages involved in this rough design are as follows:

1. As a starting point, assume that $V_{LO} < V_T$ and $V_{HI} = 0$ V so that the output signal swing ($V_{HI} - V_{LO}$) exceeds V_T.

2. The output voltage swing must also be provided by the inverter stage formed by Q1 and Q2. Since V_{HI} (inverter) $= V_{DD}$ and V_{LO} (inverter) $= V_{DD}/(1 + R_{LOAD}/R_{ON})$, then we can equate the output swing to the inverter signal swing:

$$V_{HI} - V_{LO} = V_{DD}\left(1 - \frac{1}{1 + R_{LOAD}/R_{ON}}\right) = V_{DD}\left(\frac{R_{LOAD}}{R_{ON} + R_{LOAD}}\right)$$

Hence,

$$\frac{V_{DD}}{V_T} > \frac{R_{ON} + R_{LOAD}}{R_{LOAD}} > 1 + \frac{R_{ON}}{R_{LOAD}}$$

Since we assumed that $V_{HI} = 0$ V, then the gate–source voltages of Q1 and Q2 will be identical (equal to zero) in this condition, and we can say that

$$\frac{R_{LOAD}}{R_{ON}} = \frac{\text{width of Q1}}{\text{width of Q2}}$$

and hence,

$$\frac{V_{DD}}{V_T} > 1 + \frac{\text{width of Q2}}{\text{width of Q1}}$$

provided that the inverter logic low voltage does not take the switch device out of the saturation region ($V_{DS} > (V_{GS} - V_T)$). This determines the ratio of Q1 and Q2 sizes. A ratio of 2 : 1 or 1 : 1 is typical for typical values of V_{DD} in the range 2–3 V. Increasing the width of Q1 will give some degree of margin to ensure that $V_{LO} < V_T$.

3. From the logic low conditions for the inverter and the level-shift output, calculate the number of diodes necessary to give the required level-shift assuming a forward voltage drop of 0.7 V per diode.

4. From the number of level-shift diodes, calculate the voltage at the top of the diode chain consistent with the (assumed) initial value of zero volts for the logic high output. V_{DD} must be at least equal to this level, and should preferably exceed it by a margin sufficient to keep Q3 in saturation. If the value of V_{DD} assumed at Step 2 does not meet this criterion, then Steps 2 to 4 should be reiterated. V_{DD} should not, however, be made too large since this will directly affect the gate power dissipation.

5. For correct operation of the level-shift stage, Q4 must be in saturation, i.e. $V_{DS} > (V_{GS} - V_T)$. This condition and the minimum output voltage (V_{LO}) can be used to determine the most positive level allowable for the negative supply rail V_{SS}. Again, some safety margin (e.g. 0.5 V) is wise practice, but a too negative V_{SS} will merely increase the gate power dissipation without added benefit.

6. The sizing of Q3 and Q4 in the level-shift stage determines the drive capability of the logic gate, and thus the dependence of the gate delay on loading by other gates and stray capacitance. Typically, the widths of Q3 and Q4 are approximately four times that of Q2, but determination of the optimum ratio will demand detailed simulations of the entire logic design. The gate delay can be decreased by increasing the width of Q3 and Q4, but only at the expense of increased power dissipation. Large widths should thus be restricted to those parts of the circuit where timing is critical.

7. Series resistance in the diodes can reduce the voltage swing available at the output relative to the inverter stage output. This resistance should thus be minimized by appropriate sizing of these diodes; the series resistance is generally inversely proportional to the diode area, although the aspect ratio also has an impact (see Chapter 6). Diodes with the same gate area as Q3 and Q4 can be used, but more detailed calculations may also need to be carried out to ensure the voltage swing is not severely degraded.

In comparison with DCFL, BFL has a much higher power dissipation due to the extra level-shifting stage and the requirement for dual voltage rails. In return though, the buffering provided by the level-shift stage generally gives a lower propagation delay (particularly where the capacitance load is high), and the processing is easier (only D-MESFETs are required) with a much better tolerance to variations in the MESFET characteristics.

3.4 Schottky diode FET logic (SDFL) [28]

If the power dissipation of BFL is too high for a required application, an alternative configuration exists which reduces the power dissipation by moving the level-shifting function from the logic gate output to the input of the following stage. Since the output from the bottom of the level-shift diode chain now only has to drive the input of one logic gate, less drive capability is required and much smaller devices (or devices with an increased gate length) can be used; the power dissipation is then reduced correspondingly. At low fan-outs the buffering performed by the source–follower MESFET at the top of the BFL diode chain is also unnecessary so this device can be omitted thus reducing the component count. In this case the load (Q2) should be scaled so that it is large enough to source the current required by the level-shift chains at the inputs of the driven gates; typically a width ratio of 3 : 1 is used for Q2 : Q3. The resulting circuit is shown in Figure 3.8; logic designed using this configuration is known as Schottky diode FET logic or SDFL. An approximate design for this logic can be carried out based on the previously outlined schedule for BFL. Again, detailed analysis should be carried out so that the noise margin of the gate can be optimized.

A comparison of the properties of BFL, SDFL and DCFL GaAs logic circuits fabricated with a one micron gate length is made in Table 3.2. The bracketed figures in this table emphasize that these values only indicate the order of magnitude to be expected, not absolute values. In comparing the processing tolerance, this column indicates the relative insensitivity to variations in the threshold voltage. Both BFL and SDFL are relatively good in this respect. DCFL, however, requires two types of device to be fabricated on the same IC, and furthermore the threshold voltage of the switch must be quite tightly controlled. Further discussion of this processing tolerance is made in Section 3.7 on noise margins.

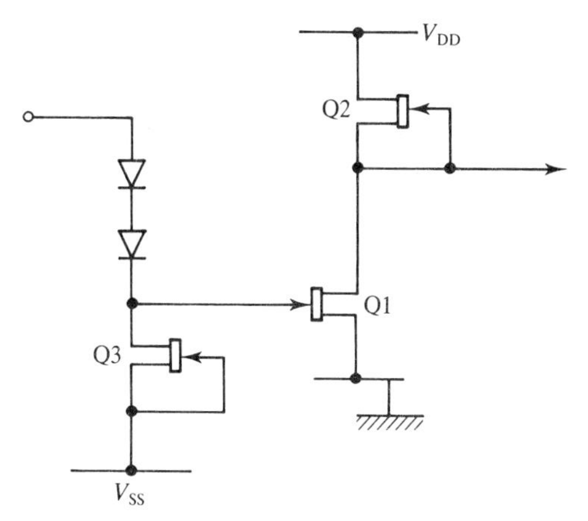

Figure 3.8 A Schottky diode FET logic (SDFL) inverter (typical parameters: $V_{DD} = 3$ V, $V_{SS} = -2$ V, $W_{Q1} = 30$ μm, $W_{Q2} = 15$ μm, $W_{Q3} = 5$ μm)

Table 3.2 Comparison of GaAs logic families

Logic family	Inverter propagation delay	Inverter power dissipation	Number of components per inverter	Processing tolerance
BFL	Very fast (50 ps)	High (5 mW)	7	Good
SDFL	Medium fast (80 ps)	Medium (1 mW)	5	Good
DCFL	Fast (100 ps)	Low (0.5 mW)	2	Poor

3.5 Source-coupled FET logic (SCFL) [29]

Although the inverter structure shown in Figure 3.1 is the most common architecture used by GaAs inverters, it is not the only form possible. An alternative is shown in Figure 3.9. This circuit operates by 'steering' a fixed current through one of a pair of switches, and then using this current to develop a voltage drop across one of a pair of load devices. In this architecture the output voltage is dependent upon the size of the current being steered and the load resistance only (provided the steering switches are sufficiently wide that they do not limit the current). In contrast to the ratioed-logic families (such as BFL, SDFL and DCFL) where the current flowing through the circuit is modulated by the input signal level, in this architecture the current is not modulated but steered to different loads dependent on the difference between the input signals. This is also the principal upon which silicon emitter-coupled logic (ECL) operates.

In order to steer the current in this inverter circuit, complementary input signals are required so that the switches operate in anti-phase. If a single input only is available then

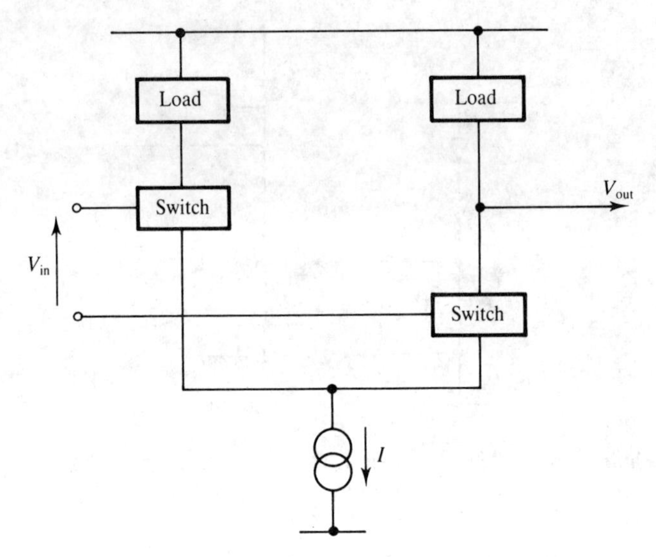

Figure 3.9 The inverter structure for current-steering logic (differential inputs)

the inverse of the signal must be generated for use as an input in order to generate the inverse as an output. This paradox can be resolved by using the inverted output to provide the required complementary input as shown in Figure 3.10.

The implementation of this architecture with GaAs MESFETs is known as source-coupled FET logic (SCFL). The current is supplied by a D-MESFET with its gate and source tied together (i.e. $V_{GS} = 0$), while the loads may be either D-MESFETs or resistors. D-MESFET loads offer the same advantages of tracking, area and speed described for the ratioed inverter. The width ratio of the current–source and the load should be such

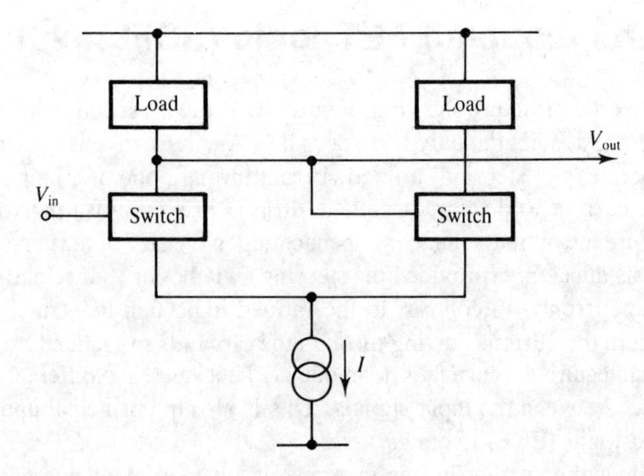

Figure 3.10 A non-differential input current-steering inverter

that the load develops sufficient voltage to be able to turn a driven switch device on and off successfully. Usually equal size devices will be used (minimum-sized to reduce the power dissipation): the current which can flow through the load has a maximum limit (I_{DSS}) which is dependent upon its width, and if an appreciably wider current source device is used, the excess current will be drawn in through the input of the switch device (see Figure 3.11 and Chapter 4). This will lower the input impedance undesirably. Conversely, if a load is appreciably wider than the current source, then the reduced load resistance will reduce the output swing unnecessarily.

The devices used for the switches may be either D-MESFETs or E-MESFETs. As was noted above, these devices should be of sufficient width that they do not limit the current flowing to the load devices. Unlike the ratioed logic families, however, the actual size is not critical, and the current in the switches is steered according to the voltage differential between the two inputs rather than being controlled by the absolute level. Thus the circuitry required is identical whether E-MESFET or D-MESFET switches are used, and extra level-shifting circuitry does not need to be included when D-MESFET switches are used, as is the case for BFL and SDFL. Having said that, however, level-shifting buffers often are included to shift the input switching level close to mid-rail.

Figure 3.12 shows the resultant circuit diagram of a D-MESFET-only inverter. Using D-MESFETs for the switches does give some advantage in that the processing will be easier than for circuits requiring both E-MESFETs and D-MESFETs. The feedback shown in Figure 3.12 is taken from the bottom of the level-shift diode chain; however, outputs are also available from higher up the chain. The circuit will also give differential outputs if required. The application of these extra outputs is described further in the section on complex gates, which occurs later in this chapter.

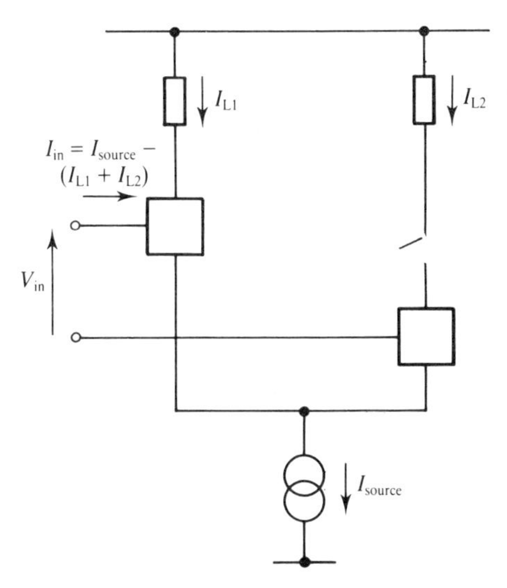

Figure 3.11 Reduced input impedance in SCFL when $W_{source} > (W_{load1} + W_{load2})$

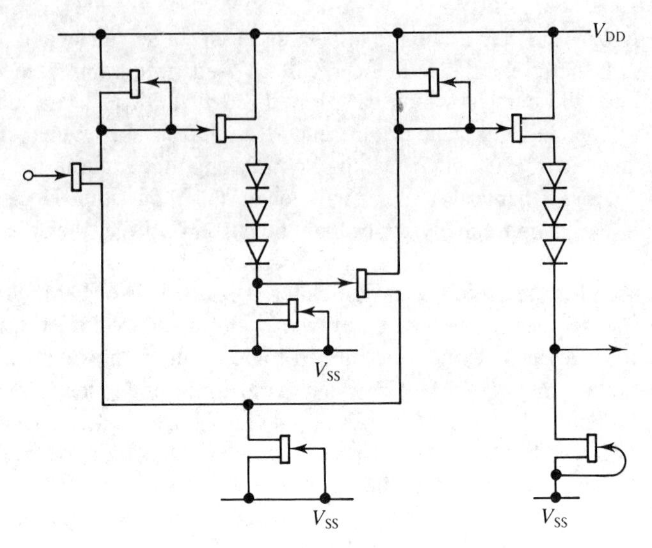

Figure 3.12 An SCFL inverter with a non-differential input (typical parameters: V_{DD} = 5 V, V_{SS} = 0 V, buffer FET widths = 20 μm, all other FET widths = 5 μm)

In comparison with the ratioed-logic families, SCFL offers a similar speed to BFL but a higher power dissipation (particularly if differential outputs, and hence two buffer stages, are required), although this power dissipation can be reduced by using a longer gate length or deliberately increasing the parasitic source resistance for the current–source device (so as to reduce the current level). The component count of the simple SCFL inverter is also high. The interest in SCFL arises from the significant advantage that complex gates can be constructed with little degradation of this speed or power, and only a small increase in the number of components. The effective power dissipation per gate and component count per gate thus falls rapidly as the gate complexity rises. This will also be discussed further later on in this chapter.

3.6 Alternative logic circuits

The emphasis of this chapter has been on the discussion of BFL, SDFL, DCFL and SCFL circuits. Other families of logic exist though, with various levels of performance. This section very briefly describes some of the other different variants available.

3.6.1 Capacitor-coupled logic (CCL) [30, 31]

This variant uses capacitive-coupling between stages to give the necessary level-shifting to drive D-MESFET-only logic. On the initial application of positive-going signals to the capacitor, the gate–source diode of the switch device will forward-bias and charge the capacitor so that at the most-positive input level the voltage on the gate is limited to

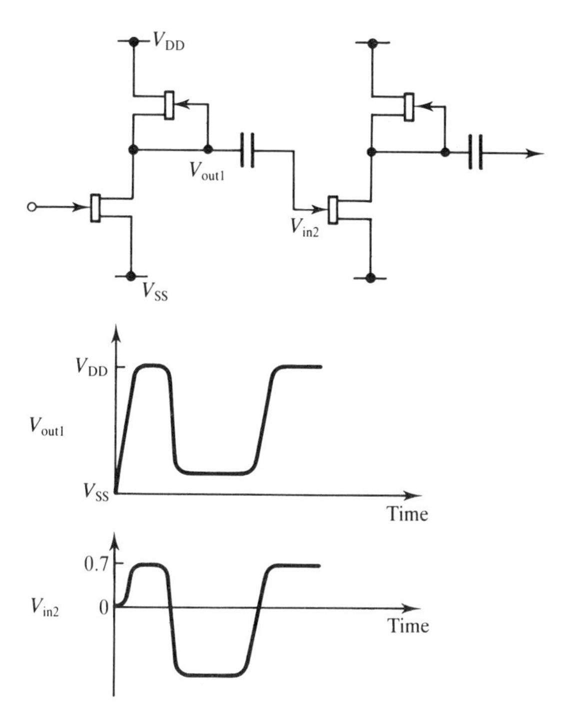

Figure 3.13 Capacitor-coupled logic (CCL)

approximately 0.7 V (see Figure 3.13). Subsequent negative-going signals can then take the gate voltage below threshold (provided a sufficiently large voltage swing is applied) so that the D-MESFET switch device can be turned off. Gate leakage currents will limit the time for which these states can be maintained (the gate bias will eventually return to zero if the positive or negative input level is held constant), which means that operation is restricted to frequencies above a minimum which is determined by the sizes of the capacitor and the switch device. However, CCL offers a lower power consumption than BFL and SDFL due to the absence of any power-consuming level-shift stage. Generally the capacitor in a CCL gate is implemented with a reversed-biased Schottky diode [32] rather than an MIM structure since this allows a reduced area; the variation in the diode capacitance with bias does not affect the gate operation.

3.6.2 Capacitor diode FET logic (CDFL) [33]

In order to overcome the problem of dynamic-only operation of the CCL family, a variant called capacitor-diode FET logic (CDFL) has been devised. Here a d.c. level-shift stage is added in parallel with the capacitor. Since this stage is only required to provide coupling at low frequencies, the drive requirements are small and thus a very low current and power dissipation can be maintained in this extra stage.

The circuit for CDFL is shown in Figure 3.14. It is of interest to note that this logic

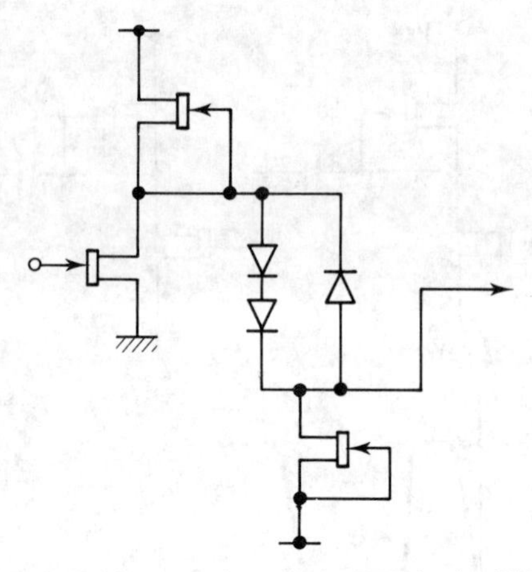

Figure 3.14 Capacitor-diode FET logic (CDFL)

family allowed one of the first commercial manufacturers of GaAs ICs (Gigabit Logic Inc.) to achieve successful penetration into the IC marketplace.

3.6.3 Super buffer FET logic (SBFL) [34]

The final variant to be described is shown in Figure 3.15. This logic is similar to DCFL, but each gate has a push–pull output stage providing increased drive capability. A push–pull output stage provides faster rise-times than a source–follower buffer, since the pull-down device (Q4) is actively turned off when the output voltage is rising, which means that all of the current from the pull-up device (Q3) is available for driving the output load; in a source–follower stage the pull-down device is always on ($V_{GS} = 0$) and this reduces the available output current. The buffer gives a better speed than DCFL under conditions of high load capacitance (i.e. high fan-out), but at the expense of increased power dissipation and gate complexity. To minimize the total power dissipation, the use of SBFL gates may

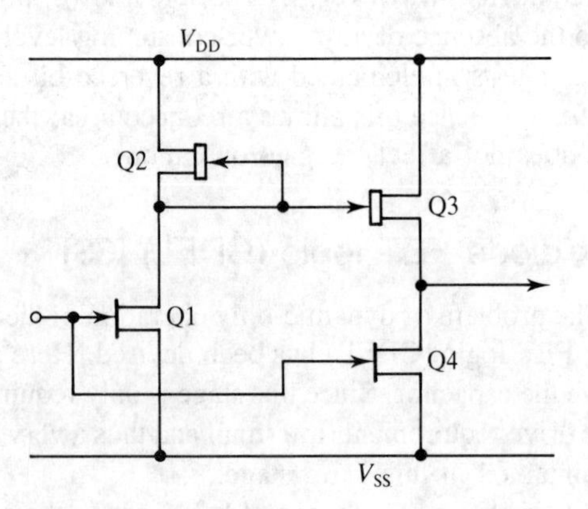

Figure 3.15 Super buffer FET logic (SBFL)

be restricted to those points within a design where the load requirements are high, for example, within the output buffers of the IC (see Section 3.12).

3.7 GaAs logic noise margins

The transfer characteristic of a circuit defines the relationship between the input and the output signals. If we measure or calculate the transfer characteristics of a single GaAs logic gate, the result is a curve of the form shown in Figure 3.16. As the input voltage is increased, appreciable current begins to flow through the switch device when the input voltage exceeds the switch threshold voltage. Further increments in the input voltage produce corresponding decrements in the output voltage as the switch resistance is reduced. At large positive values of input voltage (when the output voltage is low), the gate-drain diode of the switch device can become forward-biased, and after this point increases in the input voltage will also increase the output level. The transfer characteristic thus has a minimum as shown, this minimum being determined by the switch on-resistance and the load resistance.

Any changes in the switch threshold voltage manifest themselves by shifting the transfer curve as shown in Figure 3.16. If this change is appreciable, then the situation may be reached where input logic high or low voltages do not produce correct output logic low or high values, and in this situation the logic circuit would fail to work. The shifting of the transfer characteristic can be considered as being equivalent to superimposing a noise voltage onto the input voltage applied to a standard logic gate. The amount by which the characteristics can be shifted and still give reliable operation is thus known as the noise margin of the logic gate.

Calculation of the noise margin is most easily carried out by first considering the pair of logic gates shown in Figure 3.17. When coupled as shown, one output will sit at logic '0' and the other at logic '1'. The actual voltage levels of these outputs can be found by first plotting V_{OUT} against V_{IN} for one gate (#1), and then using the output from this gate as the input for the other to plot V_{OUT} against V_{IN} for the second gate (#2). The result

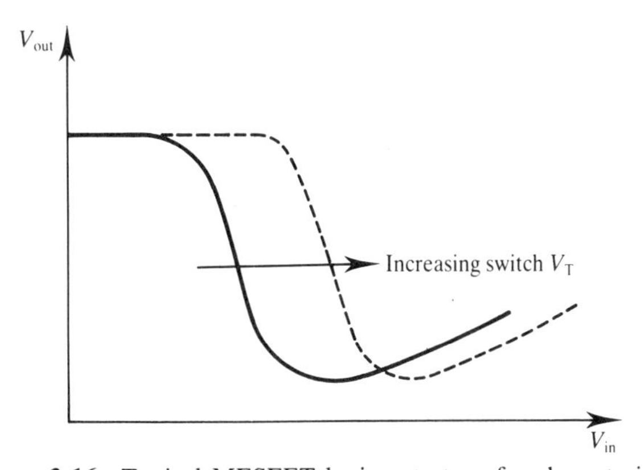

Figure 3.16 Typical MESFET logic gate transfer characteristics

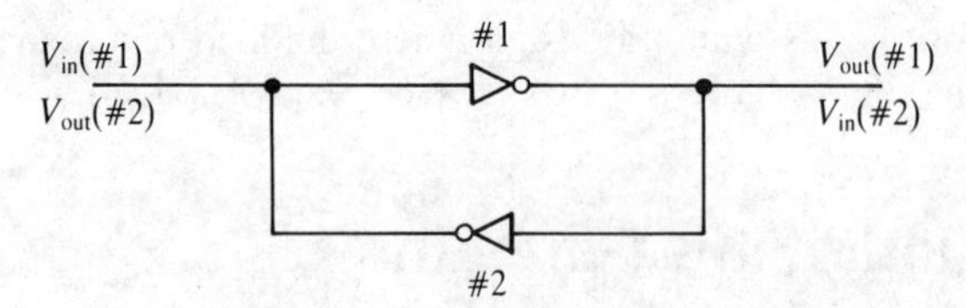

Figure 3.17 Cross-coupled logic gates for noise margin calculations

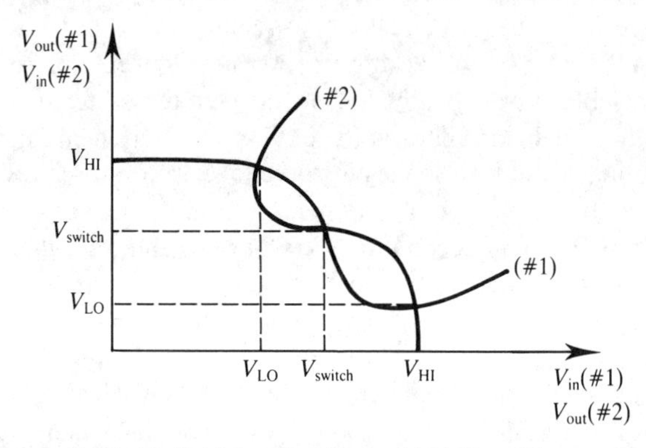

Figure 3.18 Definition of stable logic high and low voltages

is shown in Figure 3.18. The intersections of the resultant two curves show values where $V_{OUT(\#1)} = V_{IN(\#2)}$ and $V_{OUT(\#2)} = V_{IN(\#1)}$; these values correspond to the stable logic high (V_{HI}) and low (V_{LO}) levels, and to the switching point (V_{SWITCH}) of the gate. If the threshold voltage of the switch device in one of the logic gates changes, the effect will be to shift its transfer curve. If the change is excessive, the result may be as shown in Figure 3.19, where there is only one stable logic level. In this situation it will be impossible to change the state of the outputs. Any circuit built having even one logic gate with this excessive threshold voltage change will not function correctly.

The limit to how much shifting of the transfer characteristics can be tolerated (i.e. the noise margin), is given by the situations where either the switching point and V_{HI}, or the switching point and V_{LO} just merge together. In these situations, there are still two stable logic levels in the circuit, but any further shifting will reduce the number of stable levels to one. The maximum shift which can be accepted is equal to the maximum width of the areas inside the cross-coupled characteristics, and the amount of shift in these situations define a positive and a negative noise margin as illustrated in Figure 3.20. The overall noise margin for the logic gate is defined as the lesser of these two values. (NB. Often a reasonable approximation of this noise margin is obtained by the difference between the original stable logic levels (V_{HI} and V_{LO}) and the points on the transfer characteristics where the slope $= -1$, as shown in Figure 3.21.)

In any real integrated circuit, there will always be some statistical variation in the threshold voltage of the devices, and as the size of the circuit (i.e. the number of logic gates) increases, then so the probability will increase of any one gate having a threshold voltage outside the noise margin limits. There is a direct link between the size of any integrated circuit

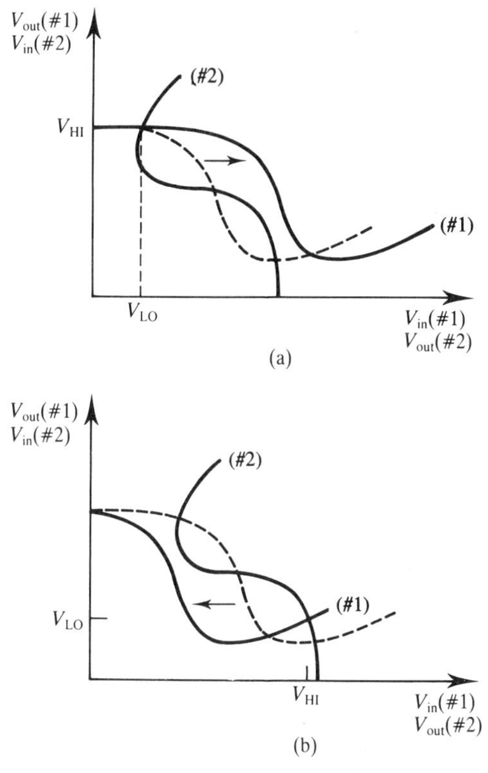

Figure 3.19 The effect of threshold voltage changes on the logic levels of cross-coupled gates: (a) excessive positive shift; (b) excessive negative shift

and the yield of working chips which is governed by the degree of statistical variation in V_T. If the mean threshold voltage of the devices in an IC is notated as V_{TM}, and the noise margin for logic gates built from these devices is notated as NM, then we can say that the IC will only function if all the logic gates have switch threshold voltages inside the range:

$$V_{TM} - NM \leq V_T \leq V_{TM} + NM$$

If the probability of one gate having a switch threshold voltage inside this range is P_{NM}, then the probability of all N logic gates in an IC having switch threshold voltages inside this range is $(P_{NM})^N$. This is the probability that the IC will function correctly. If all the (N-gate) chips on a wafer have this probability of functioning correctly, then the wafer yield will be given by:

$$\text{Yield} = (P_{NM})^N \times 100\%$$

If we assume that the threshold voltage has a normal or Gaussian distribution, then we can write [35]:

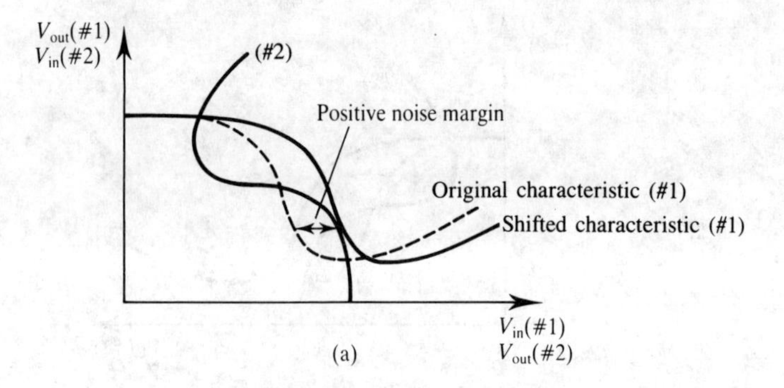

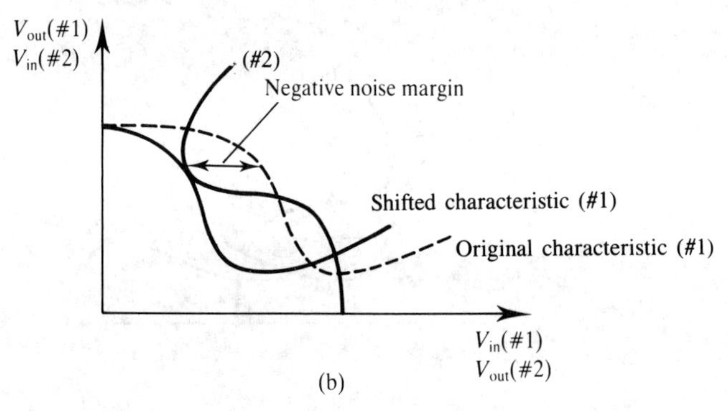

Figure 3.20 Noise margin definitions: (a) positive; (b) negative

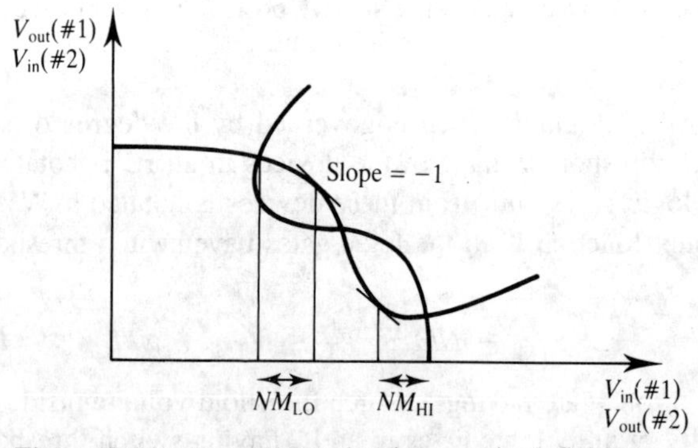

Figure 3.21 Approximated noise margin definition

$$P_{NM} = \frac{1}{\sqrt{2\pi}} \int_{-T}^{T} \exp\left(-\frac{t^2}{2}\right) \mathrm{d}t$$

where $T = NM/\sigma_{VT}$ and σ_{VT} is the standard deviation in V_T. This can be simplified to:

$$P_{NM} = \mathrm{erf}(T\sqrt{2})$$

where erf signifies the standard **error function**. Tables of error function values can be found in standard mathematical tables. The yield of an N-gate circuit will then be given by:

$$\text{Yield} = [\text{erf}(T\sqrt{2})]^N \times 100\%$$

Table 3.3 gives values of P_{NM} for various T ratios, and Figure 3.22 illustrates the variation in yield for different size circuits. It can be seen that a process giving a T value of only 3 could not be expected to produce reasonable yields for circuits of greater than 100-gate complexity. This is irrespective of any further yield degradation which might be experienced due to photolithography or packaging processing problems. It must be emphasized that the calculated values of yield shown are only due to statistical variations

Table 3.3 Variation of P_{NM} with T

T	P_{NM}
0	0
0.674 5	0.5
1	0.682 7
2	0.954 3
3	0.997 300
4	0.999 937
5	0.999 999
Infinity	1

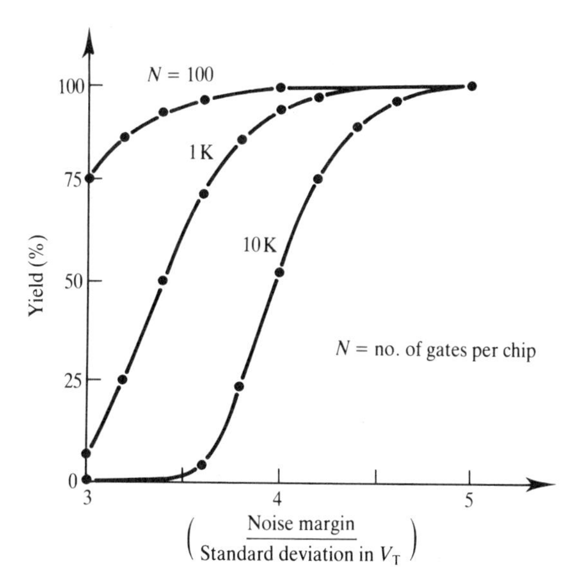

Figure 3.22 Dependence of wafer yield on chip complexity and logic gate noise margins

in V_T, no other cause of circuit failure has been considered, so these values represent an upper limit.

Having now arrived at a formula defining the dependence of yield upon the noise margin of a gate, we can now look at the characteristics of the different GaAs logic families discussed above. Both BFL and SDFL have very good noise margins, of the order of one volt or greater. This is primarily due to the large signal swings present in these circuits. Even if the standard deviation in threshold voltage were 200 mV, and in practice values well under 100 mV are easily obtained, we have:

$$T = \frac{NM}{\sigma_{VT}} = 5$$

Figure 3.22 indicates that at this level we could integrate circuits composed of 10,000 gates and still achieve yields of 99 percent (assuming perfect processing). The practical levels of integration achievable by BFL and SDFL circuits are limited by the total power dissipation of the chip, not the device uniformity. Using the previous 'order-of-magnitude' values of 3 $mW/gate^{-1}$ and 1 $mW/gate^{-1}$ for BFL and SDFL, respectively, a total power budget of 3 $W/chip^{-1}$ will limit the total circuit gate count to 1000 and 3000, respectively, for these logic families. Thus, BFL and SDFL are not suitable for VLSI integration due to the high power dissipation of each gate.

The situation for DCFL circuits is reversed. The gate dissipation is of the order of 0.3 $mW/gate^{-1}$, which limits the gate count to 10,000 at 3 $W/chip^{-1}$. However, typical noise margins are approximately 180 mV; to achieve yields of 99 percent in a 10,000 gate circuit requires a standard deviation of the switch (E-MESFET) threshold voltage of 36 mV. This uniformity requirement is appreciably harder to achieve than is the requirement for BFL or SDFL, particularly since the channel under the gate of an E-MESFET is thinner than in a D-MESFET, so that any variations in the doping density or thickness will produce a proportionally bigger change. In DCFL, the level of integration is limited by this uniformity rather than the power dissipation. Much of the emphasis of the work on GaAs DCFL VLSI logic circuits lies on improving the threshold voltage uniformity rather than developing new, high performance, circuits.

The SCFL circuits generally have a noise margin that is smaller than the DCFL noise margin (typically only 100 mV); however, SCFL operates in response to the differential voltage present between the complementary inputs, not the absolute level. The devices do not have to have a small variation in the threshold voltage throughout the whole circuit, only between each switch pair. In this situation, the probability of the circuit working is independent of the size of the circuit, and the yield will be given by the formula:

$$\text{Yield} = \text{erf}(T\sqrt{2}) \times 100\%$$

This yield value is the same as the P_{NM} values in Table 3.3. As long as the variation in threshold voltage between each switch pair is less than approximately 40 percent of the noise margin then yields of approximately 99 percent can be achieved at any order of circuit complexity. In SCFL circuits the level of integration is limited by the power dissipation which is of the same order as BFL. However, since complex gate functions can be realized in SCFL with little increase in the gate power, the limit to the level of integration is much higher than it is for BFL (this will be explained further in the following sections). SCFL thus contends with DCFL as being suitable for VLSI applications.

3.8 Complex gates

The discussion of GaAs logic has so far concentrated principally on inverter structures. In practice, the functions which can be implemented with only inverters are extremely limited! NOR or NAND gates are also required to make useful logic.

NOR gates can be simply made by adding extra switch MESFETs in parallel with the switch device in the basic inverter switch as shown in Figure 3.23. The output voltage will pull down when either of the switch MESFETs is turned on with a logic high input signal. In the case of SDFL, a further simplification can be made by wire-ORing the inputs to the inverter within the level-shift diode chain (see Figure 3.23(d)). SBFL is slightly different in that it also requires the pull-down device in the buffer stage to be paralleled; this is also shown in Figure 3.23.

NAND gates can be implemented by adding series switches to the inverter structure as shown in Figure 3.24. In this situation, the upper device is connected to the zero volt rail through the on-resistance of the lower switch; this on-resistance effectively gives the upper switch a higher parasitic source resistance which contributes to the variability in threshold voltage and thus degrades the ratio of noise margin to σ_{VT}. For BFL and SDFL type circuits, this degradation does not significantly affect the circuit yield, and NAND gates can be used provided that the width of the switch devices is increased in proportion to the number of inputs. This is required so as to maintain the same ratio of total switch on-resistance to load resistance and thus the same logic low level. Some improvement in packing density may be achieved by replacing the two series switches with a single dual-gate device. The circuit diagram and device structure in this case are shown in Figure 3.24.

SCFL operates on the input differential voltage, and this is not affected by the resistance of series switch devices, hence scaling is not necessary in this case. The supply rail voltages usually limit the number of series switches which can be used to two in the case of BFL and SDFL, and three with SCFL. The number of series switches in BFL and SDFL is also limited by the excessive area which would be required due to the width scaling which must be performed with these families. In DCFL circuits the noise margin to σ_{VT} ratio is of prime importance, and degradation of this ratio cannot be tolerated; NAND gates are thus not generally used with this logic family.

It is also possible to implement complex NOR/NAND gate structures in BFL, SDFL and SCFL. The number of series switches is again limited by the supply rails as for NAND-only structures. An example gate is shown in Figure 3.25. In BFL and SDFL the logic high and low levels are usually common for all the inputs, in SCFL correct current-steering will only be achieved if the input levels are adjusted in line with the 'level' of the input being driven. The output from an SCFL gate may thus be taken from different positions within the level-shift buffer in order to drive other gates, as shown in Figure 3.26.

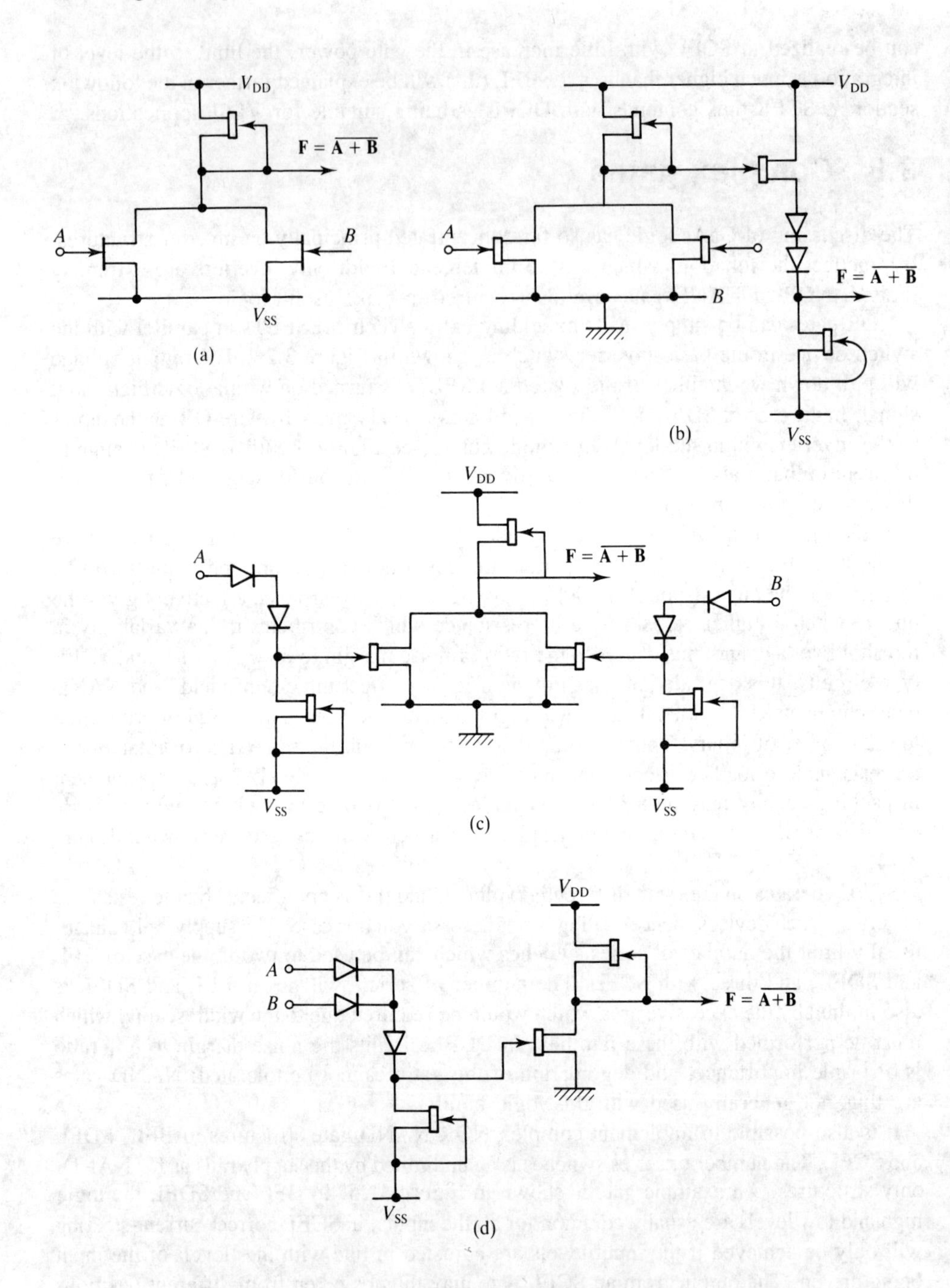

Figure 3.23 MESFET NOR gate structures: (a) DCFL; (b) BFL; (c) SDFL; (d) wire-ORed SDFL; (e) SBFL

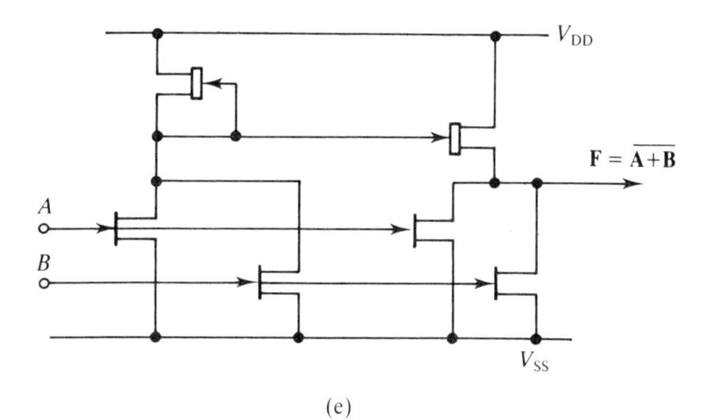

Figure 3.23 continued

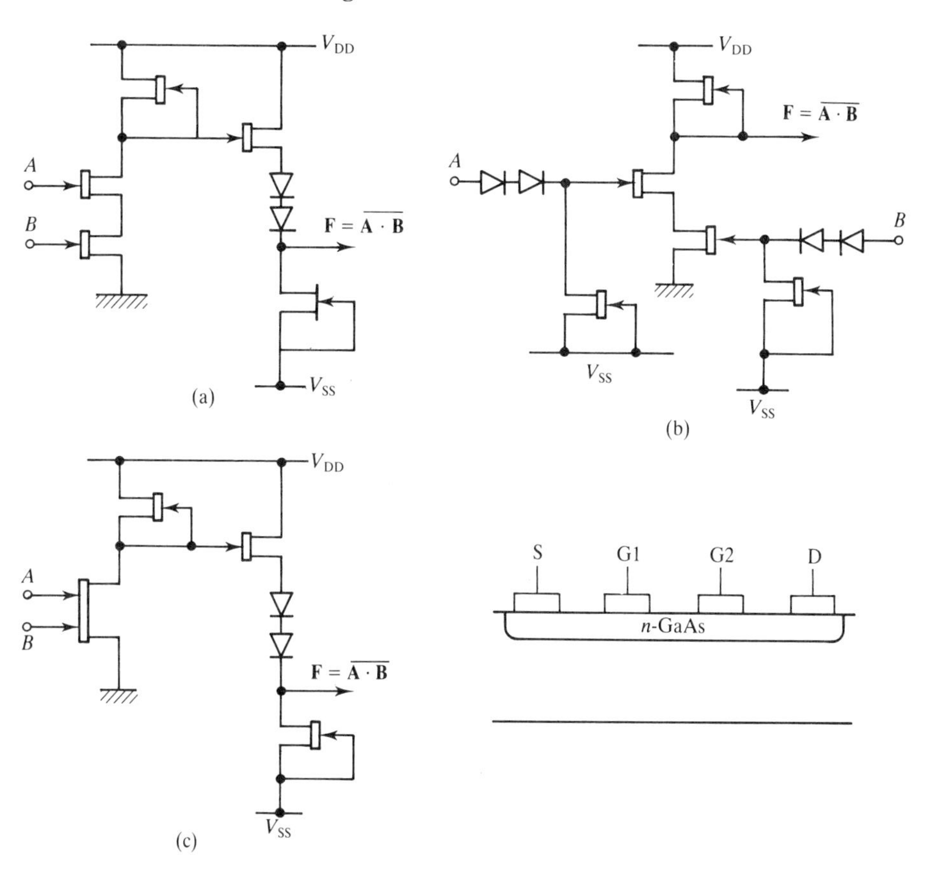

Figure 3.24 MESFET NAND gate structures: (a) BFL; (b) SDFL; (c) BFL (dual-gate input); (d) SCFL

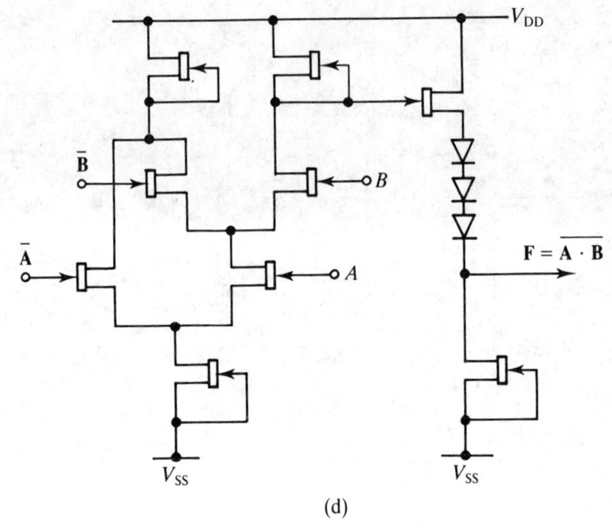

(d)

Figure 3.24 continued

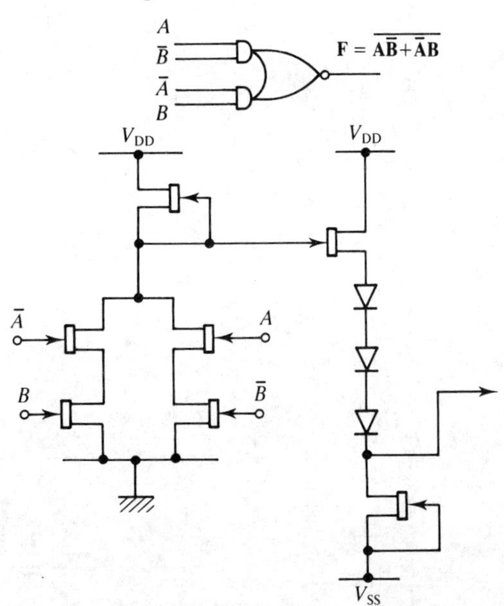

Figure 3.25 A BFL NOR/NAND gate

Complementary outputs may also be taken from SCFL gates. Complex NOR/NAND gates allow functions to be implemented with a significantly reduced number of components, lower power and faster speed compared to NOR or NAND-only implementations of the same function, and SCFL offers a significant advantage over the other GaAs families in typically allowing three levels of complex logic as well as providing complementary outputs. This is illustrated in Section 3.9.

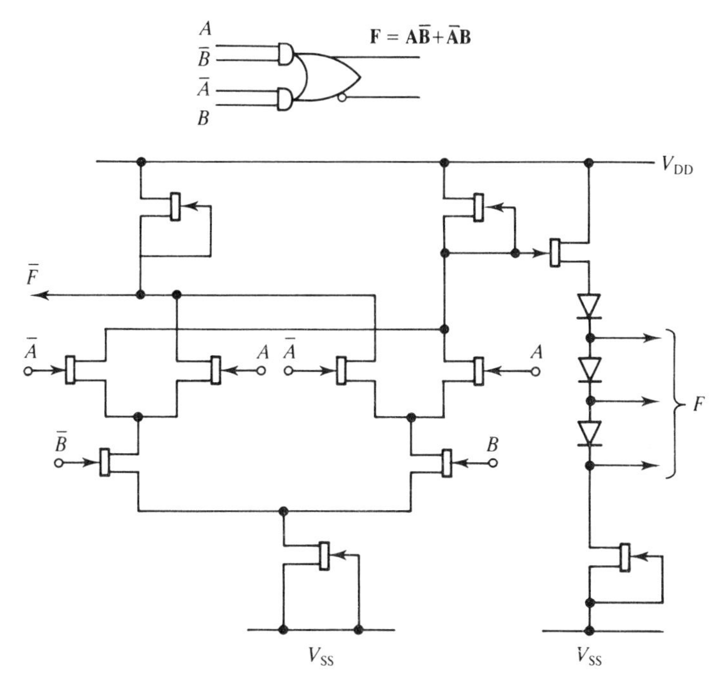

Figure 3.26 An SCFL NOR/NAND gate

The fan-in of a logic gate is defined as the number of inputs it possesses, i.e. this will be the number of switch devices. The fan-out is the number of gates being driven by the gate output. When a move is made from the basic inverter structure to more useful NOR or NAND gates, the ratioed-logic families (e.g. BFL, SDFL and DCFL) may exhibit a change in the transfer characteristics which is dependent on the fan-in and fan-out of the gate. (SCFL is not a ratioed-logic family; its characteristics are determined solely by the load and current source devices.) These changes typically take a form such that:

1. In NOR gate structures, the total on-resistance will decrease whenever more than one switch is turned on at once, and the transfer characteristic will change as shown in Figure 3.27. The logic low level decreases in these instances, which serves to increase the noise margin.
2. In NAND gate structures, the switch devices must be scaled up in width so that the total on-resistance to load resistance ratio is the same as in the basic inverter, and hence the same transfer characteristic will be obtained. If the switch devices are not scaled then the noise margin will be severely degraded as shown in Figure 3.28.
3. In SDFL and DCFL circuits, the gate inputs sink current from the driving gate (through the level-shift chain in SDFL, and into the forward-biased Schottky diode input in DCFL). As the fan-out (i.e. the number of driven gates) increases, then so will the total current increase, and the logic high level of the transfer characteristics will be

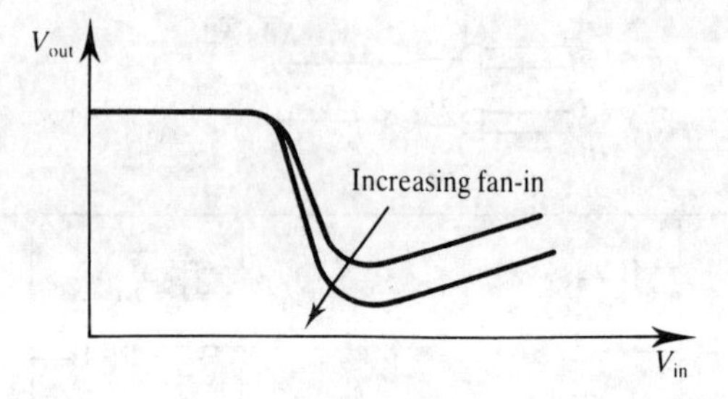

Figure 3.27 The effect of fan-in on NOR gate transfer characteristics

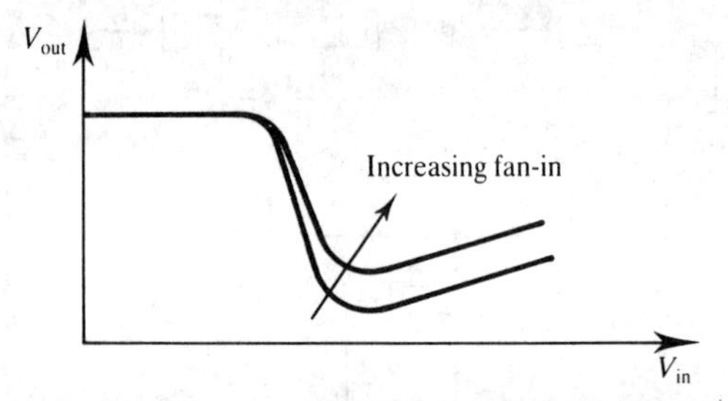

Figure 3.28 The effect of fan-in on non-scaled NAND gate transfer characteristics

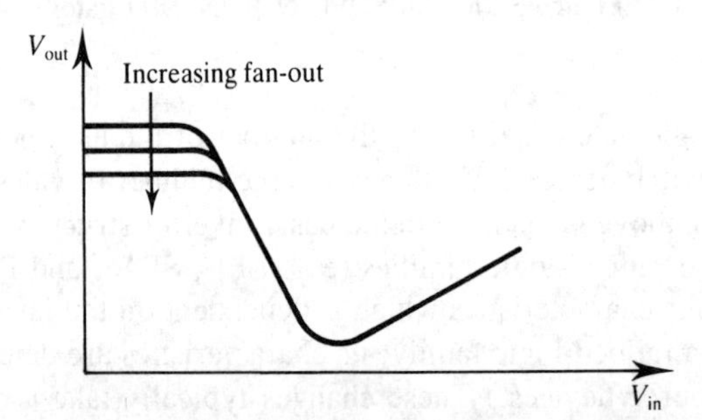

Figure 3.29 The effect of fan-out on NOR and NAND gate transfer characteristics

affected as shown in Figure 3.29. The noise margin will thus decrease with increasing fan-out. This effect may also occur with BFL gates if the output logic high is sufficiently above zero volts.

The worst-case noise margin is thus obtained under conditions of minimum fan-in and maximum fan-out, and the logic gate should be designed under these conditions. Circuit configurations which use logic gates with a fan-out higher than the maximum condition allowed for in the gate design should not be allowed. This is especially important in DCFL circuits where the noise margin is relatively small to begin with.

3.9 Logic design techniques

Although the gate design details for GaAs requires specific techniques as described above, the design methods for a complete digital IC using these gates are essentially the same as for silicon. These design methods need to be applied bearing in mind the particular constraints of the different GaAs logic families. Considering the mainstream families only, the implications of choosing a particular logic family are as follows:

1. With DCFL the small NM/σ_{VT} ratio means that NOR-only implementations should be used if reasonable yields at high levels of integration are to be achieved.
2. With BFL and SDFL then NOR, NAND and also complex NOR/NAND may be used, subject to the restriction that only two series switches may be used. The level of integration will be limited by the gate power dissipation.
3. With SCFL then NOR, NAND and complex NOR/NAND gates may be used with up to three series switches. SCFL also offers complementary outputs which may allow a reduction in the total gate count in the design. The level of integration will be limited by the gate power, although if complex gates are used this limit will be at a much higher level than BFL.

As an illustration of a design carried out using these families, Table 3.4 shows the truth table of a full-adder and Figure 3.30 the various circuit diagrams required to implement the SUM function. The circuits assume that complementary input signals are available from elsewhere in the system. Simulation of these circuits using the SPICE parameters given in Chapter 2 yields the data given in Table 3.5. This compares the relative component counts and the total MESFET widths (which affect the IC die areas), the power dissipations and the speeds of the different logic families. The large threshold voltage of the D-MESFETs necessitates a large switch to load width ratio (approximately 14) for DCFL, which gives a correspondingly large total width. Even so, this table shows that the low power, low component count and low total MESFET width make DCFL and SCFL clear contenders for VLSI applications provided that the process uniformity is sufficient to allow good yields,

Table 3.4 Truth table for a full-adder

INPUTS			OUTPUTS	
A	B	C	CARRY	SUM
0	0	0	0	0
0	0	1	0	1
0	1	0	0	1
0	1	1	1	0
1	0	0	0	1
1	0	1	1	0
1	1	0	1	0
1	1	1	1	1

Table 3.5 Comparison of BFL, DCFL and SCFL implementations of a full-adder SUM function (based on SPICE parameters given in Chapter 2)

Logic family	Component count	Total device width (μm)	Power dissipation (mW)	Delay (ps)
DCFL	17	1145	3	325
BFL	36	585	40	110
SCFL	27	285	14	70

and provided that the SCFL designs can be realized using functions which make effective use of complex gates. SCFL ICs are often realized using pre-designed standard cells to ensure this latter criterion is met.

The design process for a DCFL, BFL or SDFL IC using NOR or NAND functions is essentially based upon using De Morgan's theorems or Karnaugh maps to generate the required function; this process is well described in many textbooks [36]. The process for the design of a complex NOR/NAND SCFL gate is similar, but not so well described in the literature. For completeness, therefore, this section will conclude with an illustration of the design process for the full-adder SUM function:

$$SUM = A \cdot B \cdot C + \bar{A} \cdot \bar{B} \cdot C + \bar{A} \cdot B \cdot \bar{C} + A \cdot \bar{B} \cdot \bar{C}$$

The SCFL circuit operates by steering current through various switch devices so that a potential difference is developed across a load device in accordance with the required input signals. Thus the pair of switch devices driven by the first input signal and its complement must each be connected to another pair of switches to allow the current to

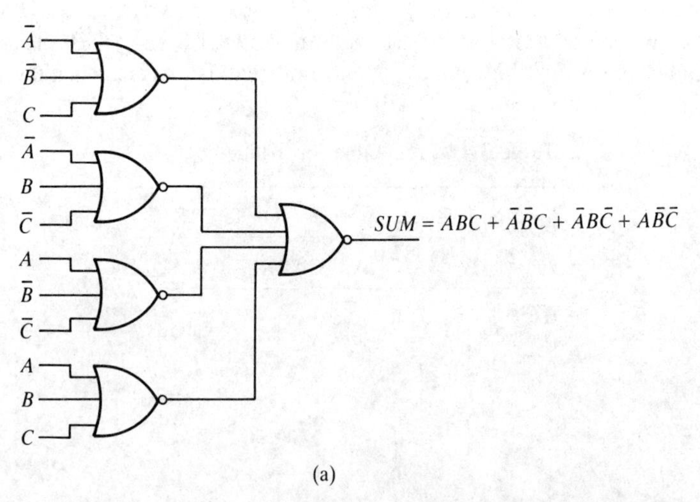

Figure 3.30 (a) A DCFL/BFL full-adder SUM circuit (see Figure 3.23 for NOR gate circuits); (b) an SCFL full-adder SUM function

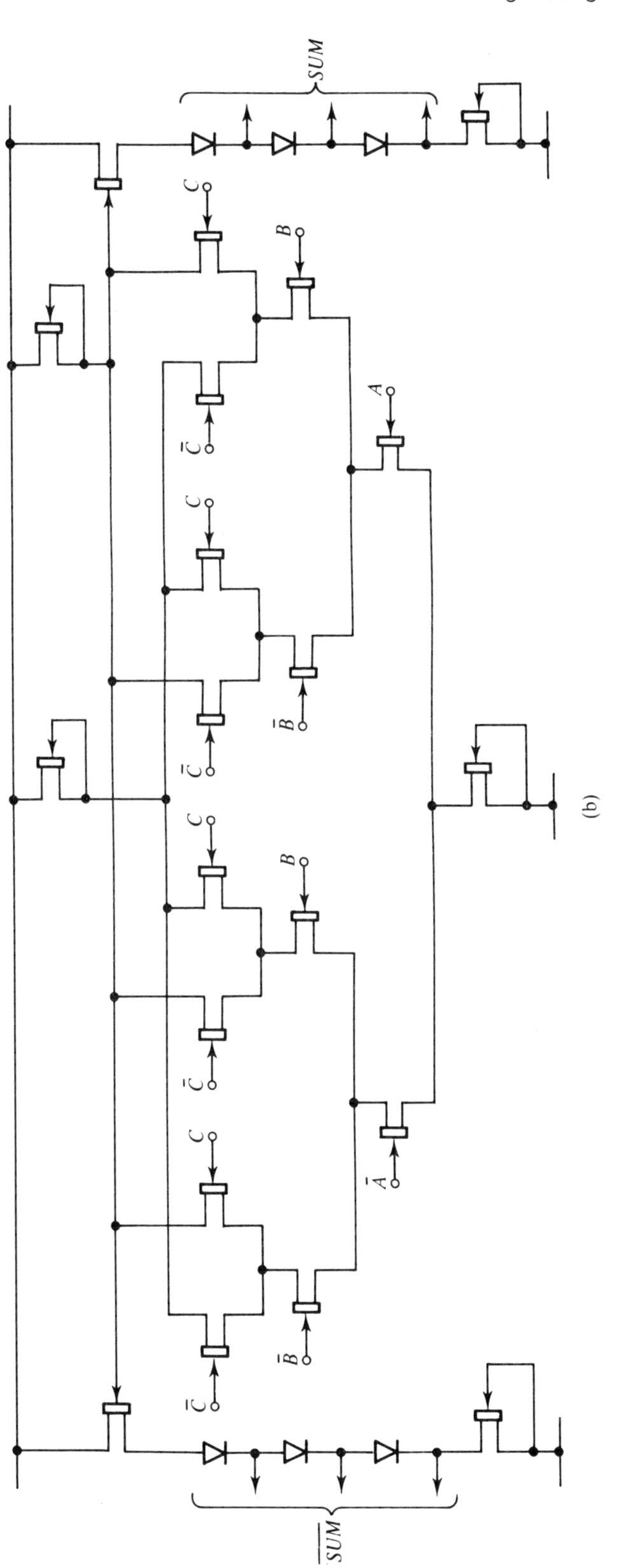

Figure 3.30 continued

be steered by the second input signal. Likewise each of the second-level switch devices must be connected to a pair of switches controlled by the third input signal and its complement (as shown in Figure 3.31). The connection of the switch devices to the loads must then be made according to the defined functionality. For the full-adder, the sum output is pulled low whenever $AB\bar{C}$, $A\bar{B}C$, $\bar{A}BC$ or $\bar{A}\bar{B}\bar{C}$ is present at the inputs. These current paths are shown in Figure 3.32. Connecting these paths together (wire-ORing) will mean that the load will develop a potential difference (i.e. the output will be pulled low) whenever current flows through any one of these paths, while remaining high under all other input conditions.

The unused current paths in this example cannot be left unterminated, but must be connected to another load device. This load provides the complement of the required function output ($\overline{SUM}$). Level-shift buffers must then be provided at the outputs to allow SUM and $\overline{SUM}$ to be used elsewhere in the system — the complete circuit diagram then becomes as shown in Figure 3.30.

In many cases it will be found that a function connects current paths which use both halves of a switch pair. In these cases the switch pair may be omitted to give a reduction in the total component count and die area. This is illustrated in Figure 3.33, which implements the CARRY function for the full-adder.

3.10 Speed considerations

For the detailed calculation of the speed performance of a circuit a CAD simulation tool (e.g. SPICE) is required. For large circuits, however, this device-level circuit simulation can be very expensive in terms of both time and computing facilities. An approximate indication of speed can be obtained using the empirical relationship:

$$\tau_D = f_i \cdot \tau_i + f_o \cdot \tau_o + l \cdot \tau_L$$

to determine the propagation delay of each GaAs gate. In this equation:

τ_i = delay per fan-in constant,
τ_o = delay per fan-out constant,
τ_L = delay per unit track length,
f_i = fan-in = number of inputs to gate,
f_o = fan-out = number of driven gates and
l = total track length connected to output.

This gate delay can be used to estimate the speed performance of the circuit either manually or through using a logic-level simulator. The third term in the expression for τ_D represents the contribution of track capacitance which in practical circuits can contribute a substantial proportion of the total gate delay. It is very important that the layout minimizes all track lengths as much as possible so that the circuit performance is not limited by this track capacitance. Determination of these parameters is described in Section 3.10.1.

If the logic gates use devices of different sizes, for example, such as when driving a large output buffer device, the definition of fan-out must be modified to:

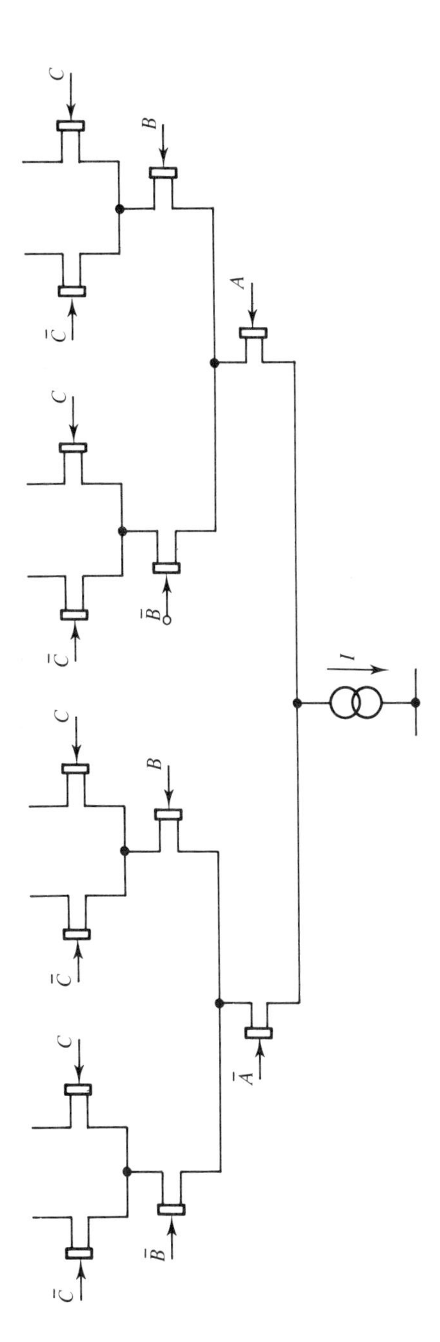

Figure 3.31 Connection of the switch devices in an SCFL complex gate

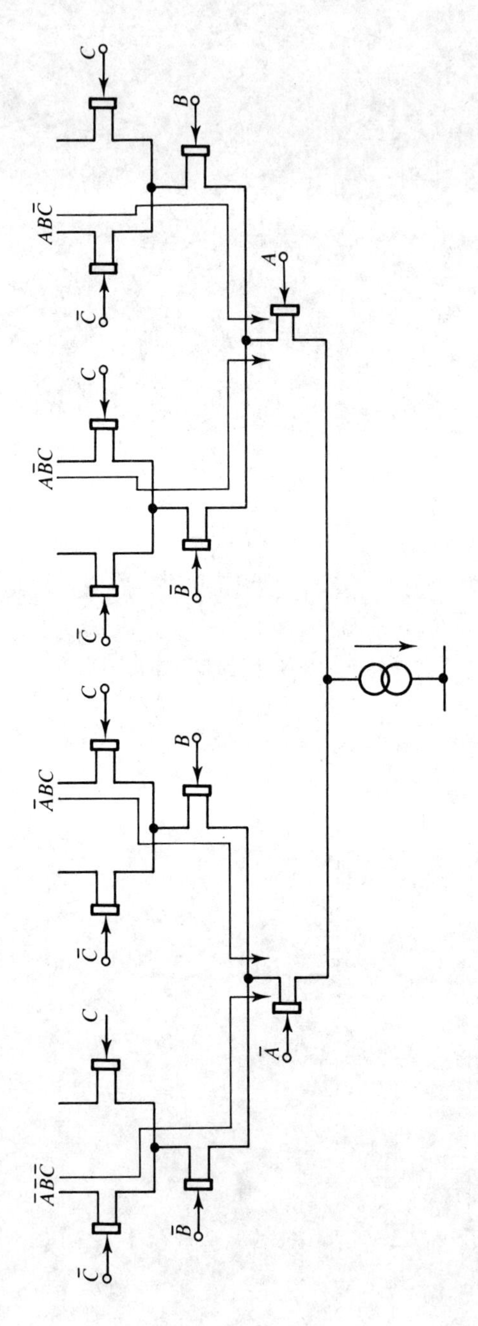

Figure 3.32 The current paths required to pull SUM low in an SCFL full-adder

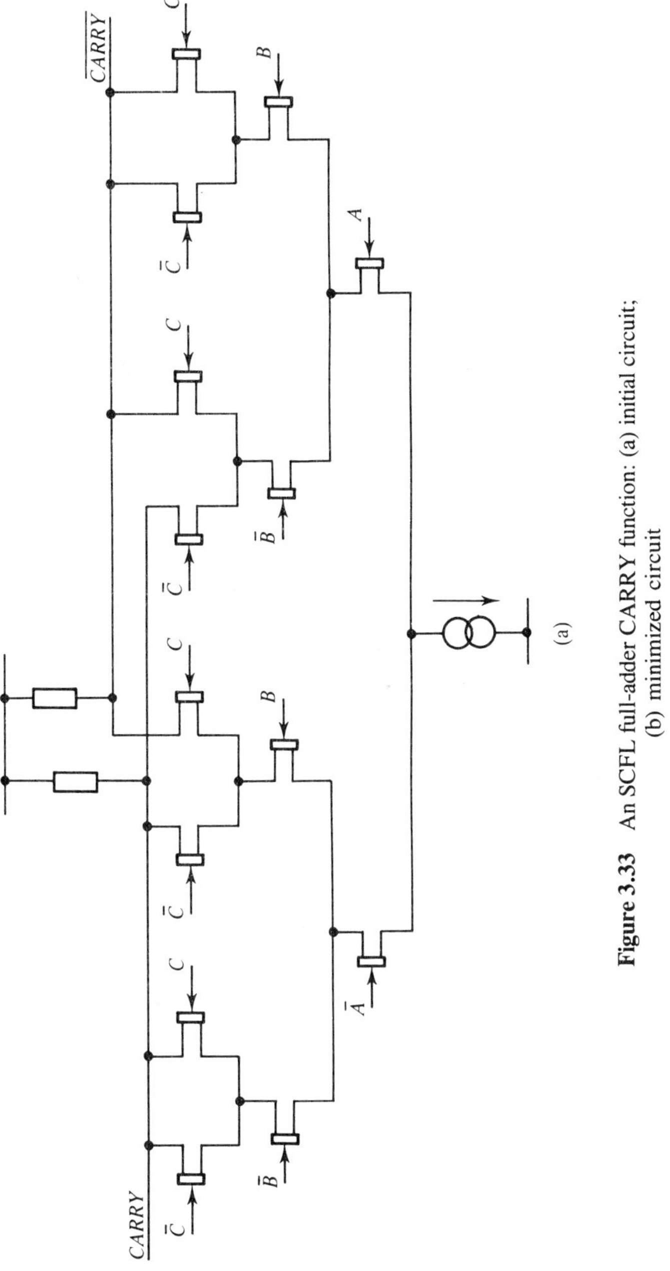

Figure 3.33 An SCFL full-adder CARRY function: (a) initial circuit; (b) minimized circuit

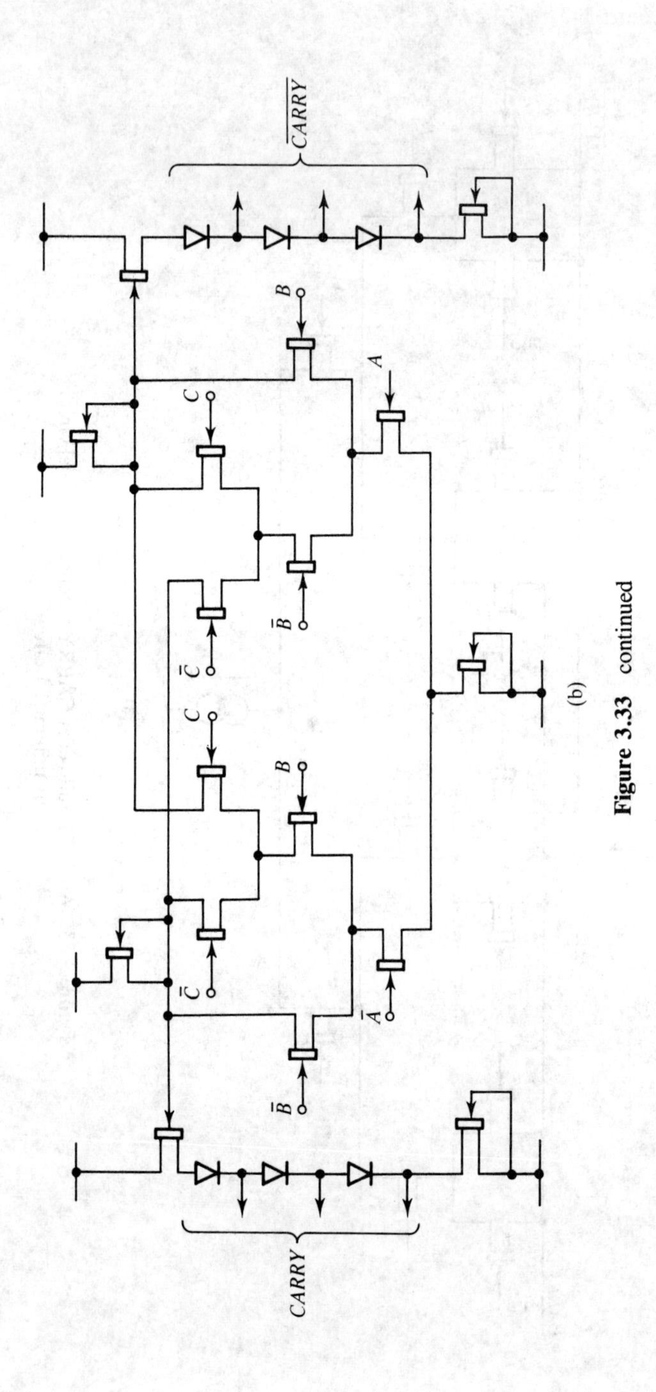

(b)

Figure 3.33 continued

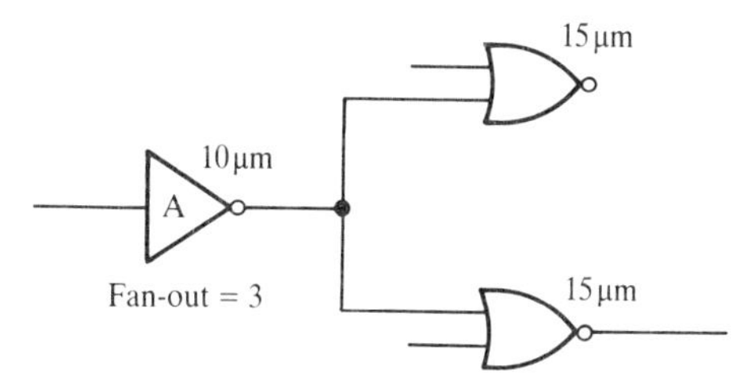

Figure 3.34 Definition of fan-out in a MESFET logic circuit

$$f_0 = \frac{\Sigma(\text{widths of driven switches})}{\text{width of driving switch}}$$

For example, in the network shown in Figure 3.34 the fan-out of gate A is 3. The values shown beside each gate in this figure indicate the width of the switch devices. The load devices will be scaled in proportion to this value in order to maintain a constant R_{LOAD}/R_{ON} ratio, and thus a constant logic low voltage level.

The variation in the fan-out which can be obtained by scaling device widths becomes important in two major situations. Firstly, if a circuit configuration has a node which is required to drive a large number of inputs (or a large single device), the driving gate can be scaled up so as to keep the fan-out within the limit imposed by noise margin considerations. Secondly, where it is important that the total delay through a chain of gates is kept small, all the gates can be scaled so as to optimize the total delay. This might be useful, for example, in the feedback network of a high-speed counter or within an output buffer stage as discussed in Section 3.12. It should be remembered, however, that there will be a change in the logic gate power dissipation associated with scaling the device widths, and this trade-off between the final circuit speed and its power dissipation should also be considered.

3.10.1 Propagation delay parameters

If an odd number of logic gates are connected in a ring as shown in Figure 3.35, the structure has no stable state and will oscillate when power is applied. A change in the output state of any gate will propagate around the ring and return to the input with a polarity which will cause a change to the opposite state to be initiated (see Figure 3.35). If the delay through each gate is τ_D, each output will thus change state every $n \cdot \tau_D$ seconds (where n is the number of stages), and will oscillate with a period given by:

$$f_{osc} = \frac{1}{2n\tau_D}$$

The gate delay can thus be calculated by measuring this oscillation frequency.

If the gate delays are measured on two ring oscillators which have different fan-in and fan-outs, then it is possible to solve two simultaneous equations for the values of τ_i and τ_o. Figure 3.36 shows suitable circuits, note that the gates loading each stage are

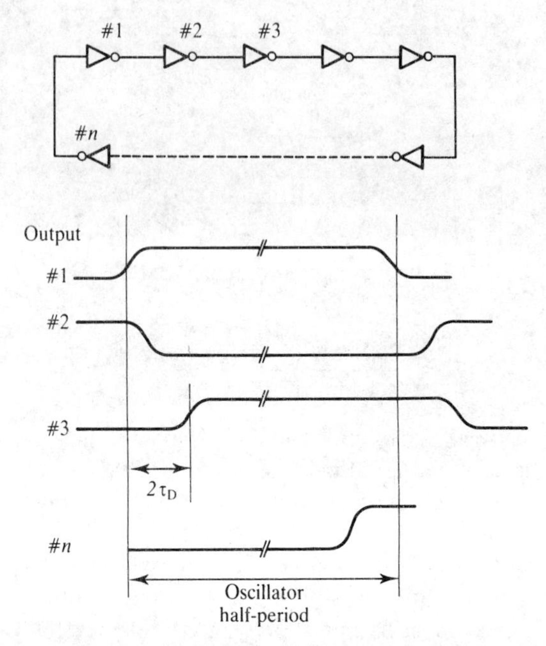

Figure 3.35 The ring oscillator structure

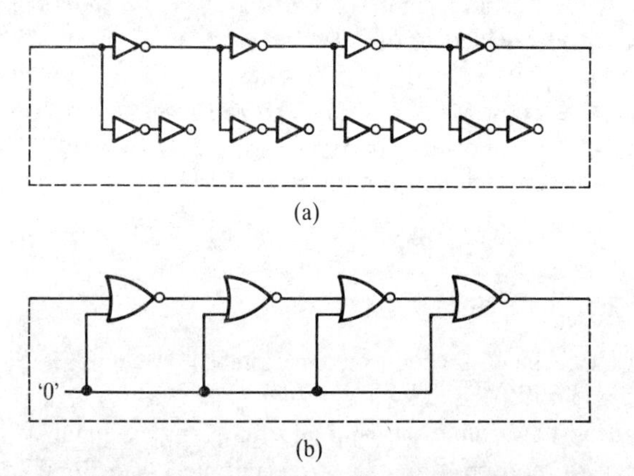

Figure 3.36 Ring oscillator structures with different fan-in and fan-out: (a) $FI = 1$, $FO = 2$; (b) $FI = 2$, $FO = 1$

themselves loaded to represent the conditions present in real circuits; omitting this loading can give misleading values of gate delay. In solving the simultaneous equations it must be assumed that the track capacitance contribution is negligible. However, ring oscillators are one of the few types of circuit where this is a reasonable assumption due to the simple nature of each stage and the restricted communication between adjacent stages only. The contribution of the track capacitance can be determined (if required), by measurements on a third ring oscillator having known track lengths connected to each gate output.

3.11 Input buffers

This chapter has so far concentrated on the details necessary to design various logic elements. The other areas which need to be discussed to enable a GaAs digital IC to be designed are the input and output buffers. These buffers must allow the IC to interface with other GaAs ICs, or with fast silicon technologies such as emitter-coupled logic (ECL) or high-speed transistor–transistor logic (TTL). This section and the next discuss GaAs IC interfaces.

The input of a digital GaAs circuit will invariably be the gate of a MESFET device. This gate is usually operated under reverse-bias conditions, or at least below the built-in voltage (approximately 0.7 V) of the diode, and the input impedance of the IC is thus very high. This high impedance makes GaAs ICs prone to damage by static discharge. Damage may also be caused by transient spikes on the supply rails. These spikes can induce currents in the input lines which are of sufficient magnitude to vaporize the metal part of the Schottky diode. Prevention of these supply spikes and precautions against static discharge are obviously important, but some form of current-limiting on the input line is also a wise precaution. One method of achieving this is shown in Figure 3.37. Typically the current-limiting D-MESFET device should be the same size as the input device so as to limit the gate current to the same level as the saturated drain current I_{DSS}. A saturated resistor of appropriate dimensions may also be used in place of this D-MESFET current limiter [33].

The inputs to a GaAs IC will generally either originate from other GaAs ICs or from silicon ECL ICs. The details of the interface requirements will depend upon the magnitude of the voltage swing and the actual levels of the signal logic high and low. ECL ICs require power rails of 0 V, −1.7 V and −5.2 V and deliver guaranteed logic levels of −0.98 V (minimum) and −1.65 V (maximum). Figure 3.38 shows a possible interface between ECL and DCFL. In this figure R_T is the transmission line termination necessary (typically 50 Ω) to ensure that fast rise and fall times can be maintained with minimal 'ringing' on the signal waveform (see Chapter 6). In this case the voltage swing is of sufficient magnitude to turn the E-MESFET switch on and off, and the interface problem becomes one of ensuring the voltage levels are correct. The solution is the trivial one of fixing the supply rail potentials for the DCFL IC to 0 V and −1.7 V, making use of the supplies required by the ECL IC. The input swing is then from approximately +50 mV to +720 mV relative to the negative rail, ideally suited to the DCFL input requirements.

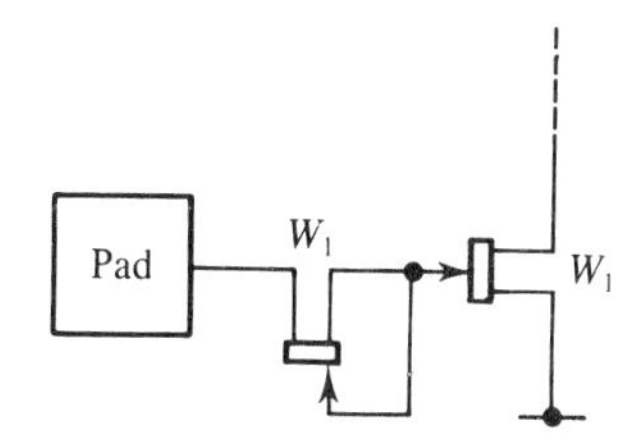

Figure 3.37 Current-limiting in input buffers

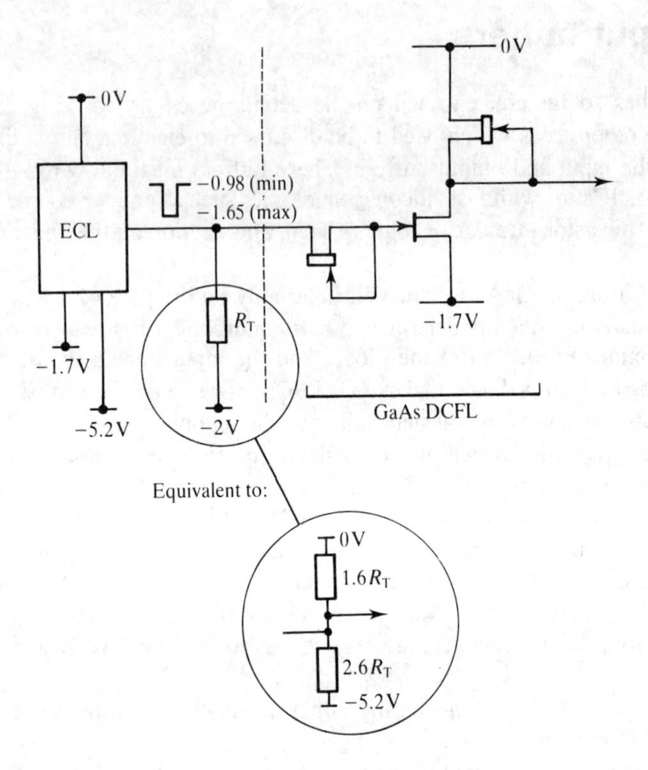

Figure 3.38 An ECL−DCFL interface

In the case of the ECL−BFL interface shown in Figure 3.39, the input BFL signal must swing the gate−source voltage of the input switch from below the threshold voltage to approximately zero volts. Typical BFL circuits will have threshold voltages in the range −1 V to −2 V, so some amplification of the ECL signal must be used to give the necessary one to two volt BFL signal swing. This amplification is provided by a differential amplifier stage, making use of the complementary outputs generally available on ECL ICs. Alternatively, one of the inputs may be driven by a constant level equal to the ECL switch-threshold voltage, this level being derived from an ECL inverter connected as shown in Figure 3.40 [37]. The amplified signal is level-shifted negatively by two diode drops (i.e. approximately 1.4 V) to provide the necessary logic high and low levels for the internal BFL circuitry.

The ECL−BFL interface shown uses six power supply rails. This can be reduced to four by firstly replacing the −2 V terminating resistor supply with an equivalent circuit supplied from 0 V and −5.2 V (as in Figure 3.38), and secondly running the BFL IC from a −5.2 V negative rail instead of −2 V, although this will result in increased power dissipation.

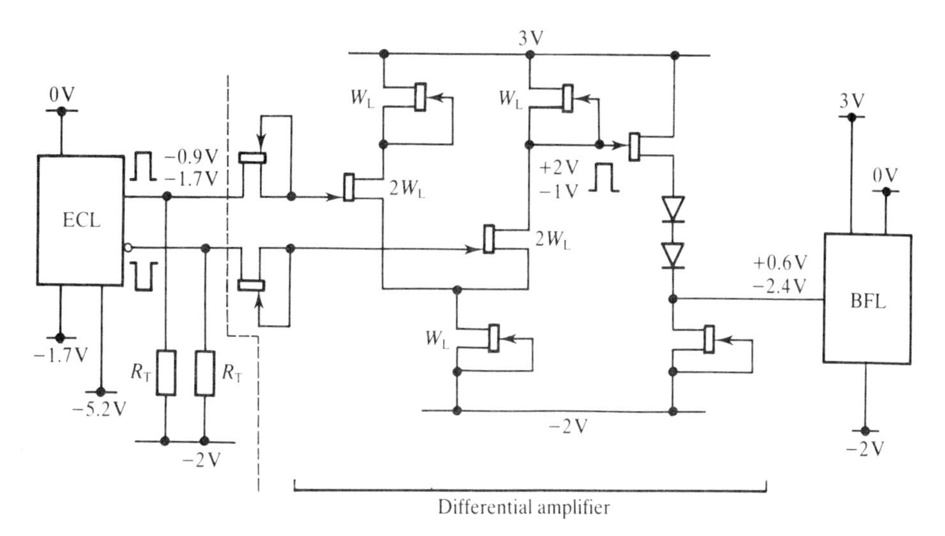

Figure 3.39 An ECL–BFL interface (differential inputs)

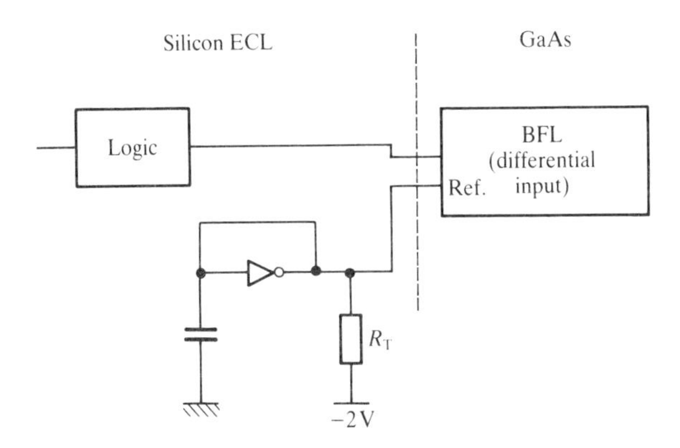

Figure 3.40 Driving BFL from ECL

3.12 Output buffers

The design problem for the output buffers is again one of achieving the correct voltage swing at the right levels for the receiving IC technology. With ECL the input levels required to ensure good noise immunity and guaranteed operation are:

$$V_{HI} > -1.0 \text{ V} \quad \text{and} \quad V_{LO} < -1.6 \text{ V}$$

while TTL requires:

$$V_{HI} > 2.4 \text{ V} \quad \text{and} \quad V_{LO} < 0.4 \text{ V}$$

The following sections describe possible interfaces between GaAs and ECL and TTL which meet these requirements.

3.12.1 DCFL to ECL

Figure 3.41 shows one possible suggestion for a DCFL to ECL interface. The output E-MESFET Q0 is connected in the open-source configuration, and must supply a current into the transmission-line termination R_T sufficient to produce the required logic high. This requirement imposes a minimum width limit to Q0, based on the current per unit width available from the device. A second width limit is also imposed by the need for a current which can charge the capacitive load on the output within the required rise-time. This current is determined by the approximate relationship:

$$I = \frac{\partial V}{\partial T} C \simeq \frac{(V_{HI} - V_{LO})}{t_{rise}} C$$

A worst-case capacitance of 10 pF is usually assumed in this calculation, based on a typical ECL input capacitance of 3 pF and typical package strays of 3 pF. Typically, Q0 may be of the order of one to two millimetres wide. Note that the open-source configuration for Q0 allows outputs to be wire-ORed together if required.

The devices between the DCFL circuitry and the output device Q0 form a 'push–pull' driver stage. Devices Q1 and Q2 are alternatively held on and off dependent upon the state of the signals at the input and output of the inverter formed by Q3 and Q4. Q0 buffers the driving circuit from the load capacitance; however, the push–pull stage will still see a fraction of the load capacitance (typically one-tenth — see Chapter 4) at its output, and will have to drive this capacitance (in parallel with the gate–drain capacitance of Q0) with the required rise and fall times. There is thus also a minimum width requirement for devices Q1 and Q2, and likewise for Q3 and Q4 which drive the push–pull stage. In cases where a very wide device is required for Q0, an additional pre-driver stage may also be required to interface the DCFL circuitry to the output driver so as to limit the

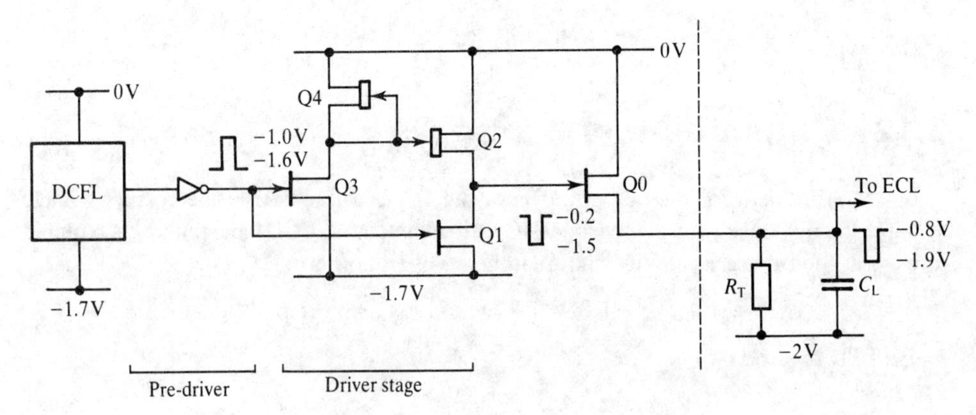

Figure 3.41 A DCFL–ECL interface

fan-out seen by each stage and to maintain fast rise and fall times. The fall time of the output signal is determined by the RC time constant of the load; if very short fall-times are required then an active pull-down device must be used. This can be implemented with another push–pull driver stage in place of the open-source output E-MESFET, using the fall-time requirement to determine the current sinking capability (and thus width) of the pull-down device.

The correct voltage levels for the DCFL–ECL interface can be simply achieved by using zero and -1.7 V rails as was suggested for the ECL–DCFL input interface.

3.12.2 BFL to ECL

Figure 3.42 shows a possible BFL–ECL interface. The output swing is provided by an open-source D-MESFET (Q0) whose width is again fixed by the requirements that it can:

(a) source sufficient current to give the required logic high voltage, and
(b) source sufficient current to maintain the required rise-time into the load capacitance.

The open-source configuration again allows wire-ORed connection of outputs if required.

The drive for the output device is based on a standard BFL stage, but with different supply rails and a different number of diodes in the level-shift section. As with the DCFL–ECL interface, this driver must also be able to charge and discharge at the required speed the fraction of the load capacitance which is seen at the input of the output D-MESFET. Extra drive capability may be conferred by the addition of the source–follower buffer stage shown in the inset in Figure 3.42. If very short fall times are required then

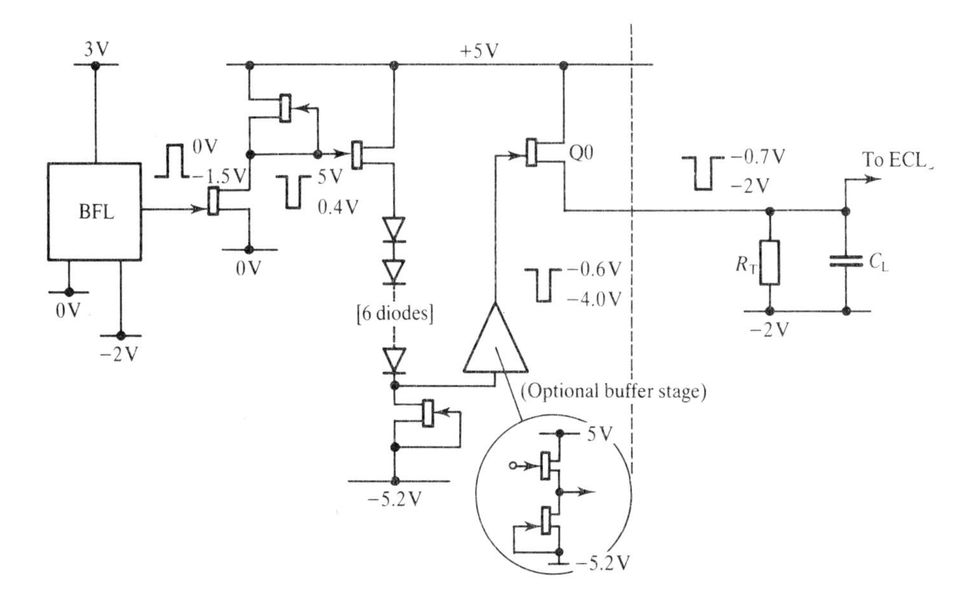

Figure 3.42 A BFL–ECL interface

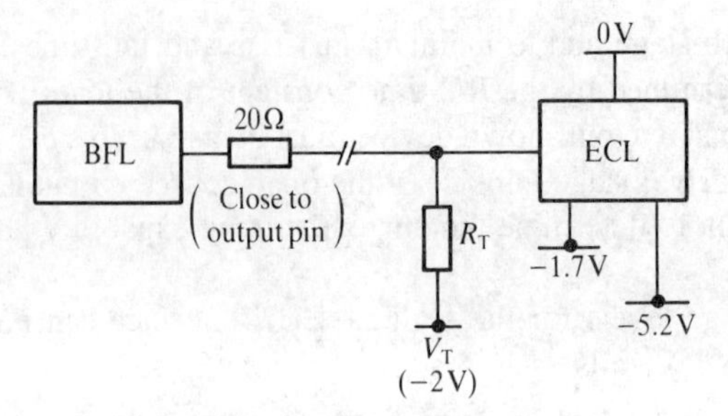

Figure 3.43 Driving ECL from BFL

an active pull-down device will again be required. A source follower buffer used in place of the open-source D-MESFET Q0 will provide this, again using the fall-time requirement to determine the pull-down device width.

In some situations the logic high voltage available from BFL may cause the input transistor in the ECL circuitry to saturate, it may then be slow to turn off when the input level is reduced. This problem can be alleviated by inserting a series 20 Ω resistor to limit the ECL input current as shown in Figure 3.43 [37]. The number of power rails required for this interface may again be reduced by replacing the −2 V terminating resistor supply with an equivalent produced from 0 V, −5.2 V supplies (see Figure 3.38), and redesigning the BFL gates so that +5 V, 0 V and −5.2 V rails can be used throughout. The IC power dissipation in this case though, will be much increased. The rails shown in Figure 3.42 (and also Figure 3.39) keep the power dissipation to a reasonable level while maintaining good speed performance and noise margins.

3.12.3 GaAs to GaAs

Although there is no international agreement to do so, the interconnection of GaAs ICs is much simplified if the ECL standard is adopted. Design of output (and input) buffers then becomes a matter of repeating the designs for ECL interfaces described above. More importantly, by adopting this standard then a GaAs IC may be simply 'dropped in' to a previously engineered ECL-based system; system designers do not have to devote time and effort getting to grips with a new input–output standard.

3.12.4 DCFL to TTL

Interfacing with TTL circuits is most easily achieved using the buffer shown in Figure 3.44. In this circuit the output device Q0 is connected in the open-drain configuration, R_T is again the transmission line termination and C_L the load capacitance. The width of Q0 is fixed by the need to:

(a) sink sufficient current to pull the output voltage down to the TTL input logic low (+0.4 V), and

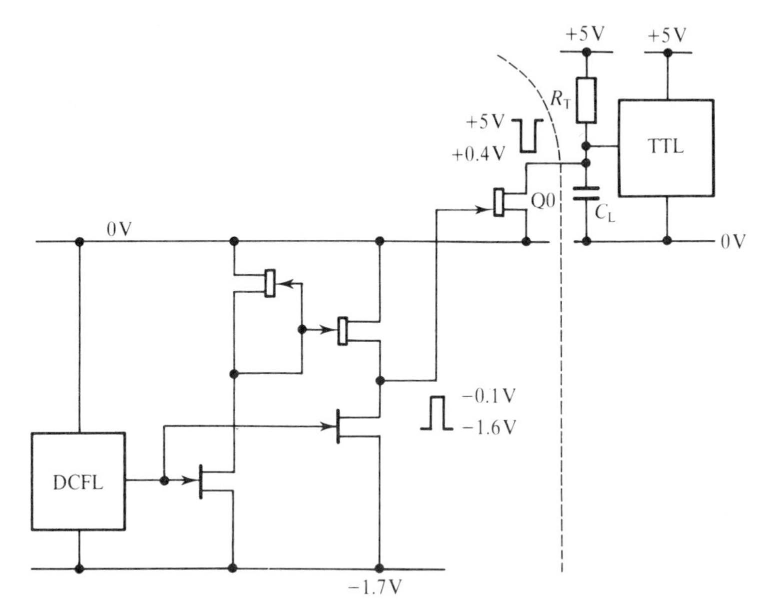

Figure 3.44 A DCFL–TTL interface

(b) sink sufficient current to discharge the load capacitance in the required fall time.

Driving of the output D-MESFET is performed by a push–pull driver as described for the DCFL–ECL interface. In this case, this driver must be capable of driving the input capacitance of Q0 which is approximately

$$C_{IN} = C_{GS} + (1 + A_V)C_{GD}$$

where C_{GS} is the D-MESFET gate–source capacitance, and C_{GD} is the gate–drain capacitance. The factor A_V is the voltage gain of the output device, and allows for the Miller multiplication of C_{GD} (see Chapter 5). A good approximation of the voltage gain is given by:

$$A_V = g_m R_T$$

for all practical values of R_T.

3.12.5 BFL to TTL

Figure 3.45 shows a possible BFL–TTL interface. The output device (Q0) is again an open-drain D-MESFET as used for the DCFL–TTL interface, but the drive is based on a standard BFL inverter stage with extra drive capability conferred by the addition of a source–follower buffer. The scaling of the device widths is determined by the same criterion described above for the DCFL interface.

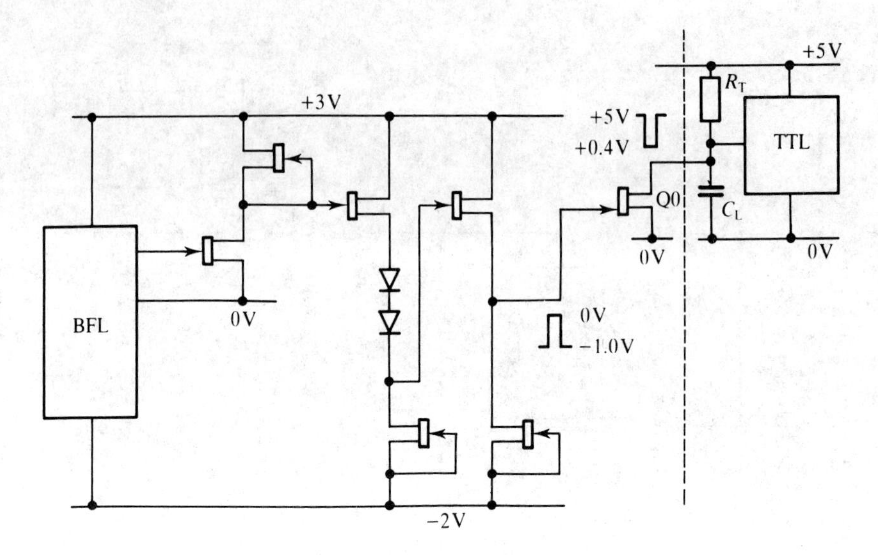

Figure 3.45 A BFL–TTL interface

4 □ Analog design

The design of analog circuits using GaAs MESFETs has many features in common with designs using silicon MOSFETs, with the important difference that it is possible to forward-bias the gate electrode; under this condition quite large gate currents may flow and the normally high input impedance of the MESFET undergoes a drastic reduction. The circuit designer must be aware of this possibility, and guard against it where necessary.

This chapter describes the properties of different building blocks which may be used to build complete GaAs analog circuits, and concludes with an example of an operational amplifier design and a high performance switched-capacitor filter built using these blocks.

4.1 Circuit analysis and d.c. biasing

As with the design of any analog (or, indeed, digital) circuit, analysis can be carried out with the assistance of a circuit simulation program (e.g. SPICE [18]). Models for GaAs devices exist in many of the commercially available CAD packages. It is not the intention of this section to compare the various attributes of these packages, suffice it to say that these packages exist to help the circuit designer.

Useful as these CAD tools undoubtedly are, a better 'feel' for the way in which a circuit operates is often gained by performing some simple 'hand' analysis. By manually deriving equations relating, say, circuit gain to the device parameters, a better understanding of the possible circuit trade-offs (e.g. gain vs bandwidth) can be gained. In order to be able to do this hand analysis, a simple model must be available. One such model which is suitable for analysing the small-signal behaviour of saturated devices is the hybrid-pi circuit shown in Figure 4.1(a). This model has been simplified in Figure 4.1(b) to allow so-called 'mid-band' analysis where the effect of the gate–source and gate–drain capacitances (C_{GS} and C_{GD}, respectively) can be assumed to be negligible. This simple model is used throughout this chapter to analyse the behaviour of various GaAs MESFET analog circuits.

In using the simple a.c. models, it should be remembered that all the parameters must

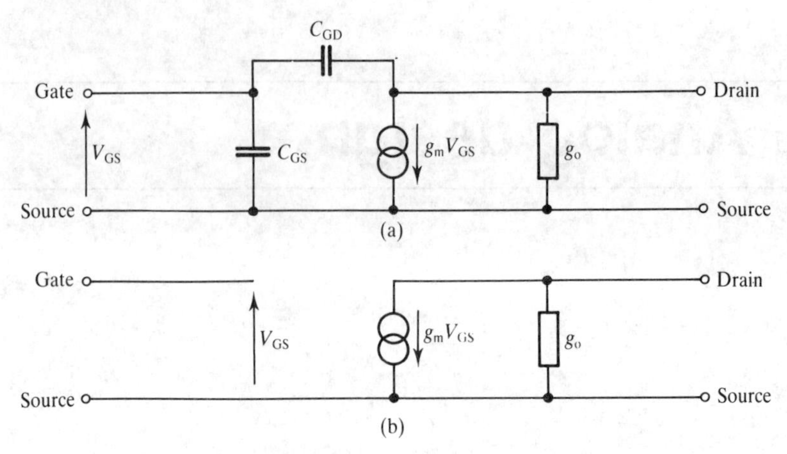

Figure 4.1 Small-signal models for a GaAs MESFET: (a) high frequency; (b) mid-band

be scaled in line with the device width (see Chapter 2). Values of C_{GS}, C_{GD}, g_m (the transconductance) and g_o (the output conductance) supplied by GaAs IC manufacturers are usually given as values scaled for a unit width device. The output conductance (g_o) is also frequency dependent, and it is very important that the correct parameter is used for the frequency of interest. Furthermore, all the parameters are bias-dependent; for example, the input impedance in the mid-band hybrid-pi model is infinite; however, this value will change drastically if the gate is forward-biased. Although the hybrid-pi model can be used in the linear region of MESFET operation, the model parameters are very different from those possessed in the saturation region. The major effect of operating in the linear region is to yield much lower values of a.c. gain than in the saturation region. The output d.c. level may also be close to one of the power rail levels if any of the devices are in the linear region, and this will restrict the maximum signal swing available from the output (i.e. the **dynamic range** of the circuit) and may introduce distortion. It is thus a general requirement that all the devices in an analog circuit be biased in the saturation region.

Achieving the correct d.c. bias conditions in a fabricated circuit may always be carried out by 'trial-and-error' — adjusting the power rails and the d.c. levels of the signals manually until all the devices are in saturation. This process can be duplicated when simulating a circuit, however, it has the severe disadvantage of being an extremely time-consuming process especially where high gain is involved. Any slight 'error' in the input signal d.c. level will be amplified so as to cause the output to saturate at one of the power rails, and the d.c. input level may need to be set with an accuracy of better than 1 mV in order to ensure correct biasing of all the devices in the circuit. A much more satisfactory approach is to apply negative feedback to the circuit at d.c. so that it automatically sets the correct bias conditions.

As an example, consider the biasing of an inverting amplifier using the feedback arrangement shown in Figure 4.2. The d.c. conditions are set by the external supply $V_{d.c.}$ feeding the resistor network R_1 and R_2. MESFET amplifiers have a high input impedance,

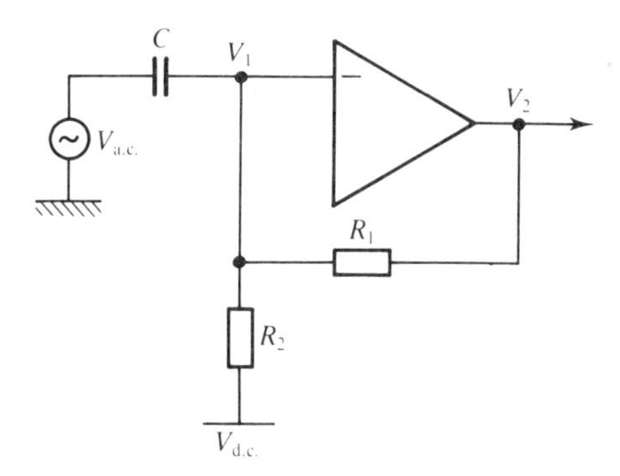

Figure 4.2 The d.c. biasing of an inverting amplifier

so the current through both resistors must be equal, and in the case when $R_1 = R_2$, we can equate the potential drops across the components by:

$$V_2 - V_1 = V_1 - V_{d.c.}$$

where V_1 and V_2 are the d.c. levels required by the internal circuitry of the amplifier to keep its devices in saturation. The required d.c. bias supply is thus:

$$V_{d.c.} = 2V_1 - V_2$$

The gain of the system for this d.c. signal is given by the ratio R_1/R_2, which is usually made equal to unity. Without the feedback, the input level (V_1) must be set very accurately to avoid saturating the output level. With the negative feedback, any error in the setting of $V_{d.c.}$ may cause V_1 to be initially incorrect, but the connection to the output (V_2) will cause V_1 to be corrected automatically. For example, if V_1 is initially too positive, this signal will be amplified, which will cause V_2 to move towards the negative rail, and this will in turn pull V_1 negatively back to the correct biasing level. The system is thus de-sensitized with respect to any slight errors in the input d.c. level, and correct biasing of the devices can be easily achieved.

The a.c. signals to this circuit are applied via a blocking capacitor, as is also shown in Figure 4.2. The gain of the system for these a.c. signals is given by the ratio of R_1 to the impedance of this capacitor. If the capacitor impedance is much less than the impedance of R_1 (and both resistors will normally be made large, i.e. greater than 1 megaohm, to avoid loading the amplifier output), then the a.c. gain will be limited to that of the amplifier alone, and the 'open loop' properties of the amplifier can then be investigated.

The transition between the 'closed loop' behaviour used to set the d.c. bias conditions and the 'open loop' a.c. behaviour occurs at the frequency where the capacitor impedance and the impedance of R_1 become equal. This frequency marks a low frequency cut-off in the system characteristics as shown in Figure 4.3. The value of C thus needs to be large enough so that this cut-off frequency is below the frequencies at which the amplifier analysis is required.

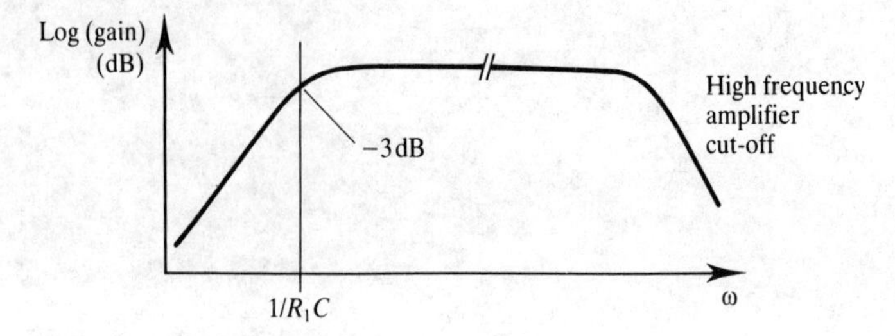

Figure 4.3 Frequency response of the amplifier shown in Figure 4.2

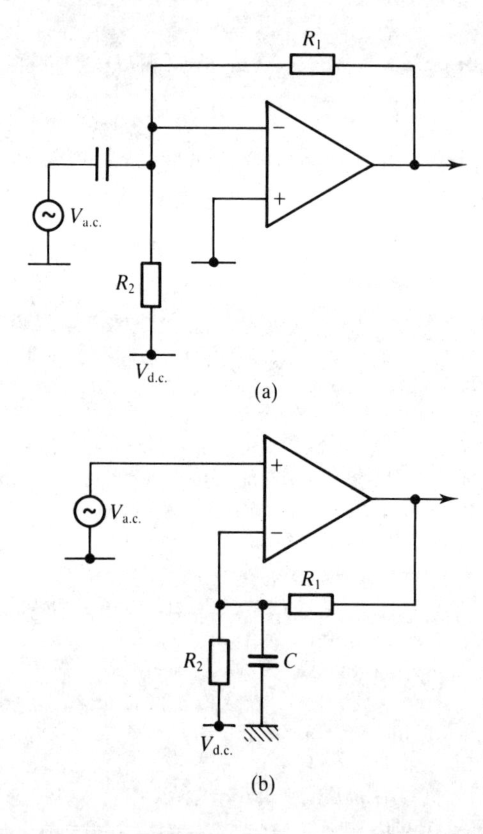

Figure 4.4 The d.c. biasing of a differential amplifier: (a) inverting input; (b) non-inverting input

For systems involving high gain differential amplifiers, a similar feedback scheme can be used to set the appropriate d.c. biasing. In this case two alternatives exist, depending upon whether the a.c. input signal is required to drive the non-inverting or inverting input. These are shown in Figure 4.4. Again, the negative feedback de-sensitizes the system from the need to set the d.c. conditions exactly, while allowing the open-loop a.c. properties of the circuit to be investigated provided that the low frequency cut-off defined by $R_1 C$ is set sufficiently low.

4.2 Analog building blocks

In the following discussion it is assumed that all the MESFETs used are depletion-mode. This is preferred in practice since it simplifies the processing schedule and gives relatively good matching between devices, and it is also relatively easy to arrange suitable bias conditions for the devices. Enhancement-mode devices may also be used if desired; however, since the gate–source bias must be positive for channel current to flow in E-MESFET devices greater care must be taken with the d.c. biasing in the circuit to ensure that undesirable effects due to gate conduction do not occur; the gate bias must be sufficient to allow a channel current to flow, but not high enough to cause forward-biasing of the gate–channel diode. The a.c. analysis of circuits using these E-MESFETs is not described in this chapter, but will follow closely the D-MESFET analysis which is given in the following sections.

4.2.1 Inverting amplifiers

Inverting amplifiers can be simply constructed by the series connection of a MESFET and a load device. Figure 4.5 shows two such amplifiers, one using an active D-MESFET load, the other a passive resistor load. If the power supply voltage is sufficient to ensure that both devices are in the saturation region ($V_{DS} > V_{GS} - V_T$), then small-signal analysis can be carried out using the simplified hybrid-pi model described above. The small-signal equivalent circuit for the inverting amplifier is shown in Figure 4.6; in the case with a D-MESFET load the gate–source voltage is fixed to zero, so the model simplifies to comprise a conductance (g_L) only which connects between the output node and V_{DD} which can be considered as an a.c. ground. The output voltage can then be expressed as:

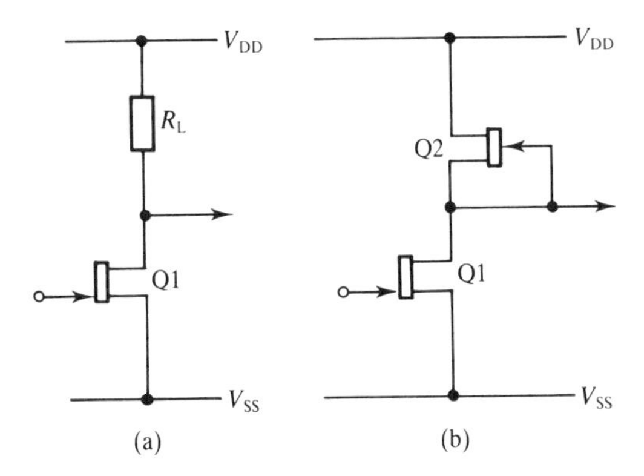

Figure 4.5 Inverting amplifiers: (a) passive load; (b) active load

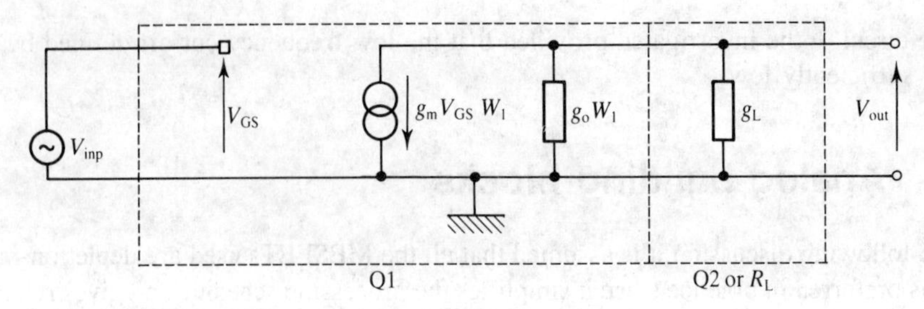

Figure 4.6 The small-signal equivalent circuit for an inverting amplifier

$$v_{out} = -\frac{g_m v_{inp} W_1}{(g_o W_1 + g_L)}$$

where g_L is the conductance of the load and $(g_m W_1)$ and $(g_o W_1)$ are the scaled parameters for the input device (Q1). This results in an a.c. voltage gain given by:

$$A_v = \frac{v_{out}}{v_{inp}} = -\frac{g_m W_1}{(g_o W_1 + g_L)}$$

Note that the negative sign shows that the input and output signals are in anti-phase, i.e. the amplifier is inverting.

In most situations the output d.c. bias is required to be mid-rail to maximize the possible output swing, which means that for the amplifier with a D-MESFET load equal size devices must be used. This criterion distinguishes analog inverters from logic inverters where the aim is to maximize the noise margin and thus unequally sized devices are used. The a.c. voltage gain then simplifies to:

$$A_v = -\frac{1}{2}\frac{g_m}{g_o}$$

(since $g_L = g_o W_1$ in this case). Typically g_m/g_o is in the range 15 to 25, which results in inverter gains of 7.5 to 12.5.

The choice of whether to use an active or a passive load (i.e. a D-MESFET or a resistor) will depend on the overall circuit specification. The active load configuration gives a higher gain in most circumstances, but the output voltage positive swing is limited by the requirement that the D-MESFET load must stay in saturation which means that it cannot swing to within V_T of the positive rail V_{DD}. The passive load will allow output swings up to the V_{DD} rail, albeit at lower gains.

If any variation in the component values occurs (e.g. due to V_T variations or processing tolerances), then the output voltage may not sit at the required d.c. bias. To compensate for this variation, a bias may be applied at the inverter input; this is the input offset voltage of the stage. Generally, the active load configuration is also preferred due to the better matching which can be obtained between the components and the consequent reduced input offset voltage. It should be noted that large positive values of input offset may forward-bias the gate of the input device, so reducing the input impedance and loading the driving stage.

4.2.2 The source–follower

High gain in analog amplifiers is frequently accompanied by high impedance. Furthermore, such high gain circuits are frequently required to drive loads of high capacitance or low resistance (i.e. low impedance). Some form of buffering is thus required between the high impedance amplifier and the low impedance load if the high gain characteristic is to be maintained; this buffering is typically performed by a source–follower circuit as shown in Figure 4.7. Q2 in this circuit acts as a constant current source (with an output conductance governed by g_o); since the current defined by this device flows through Q1, and Q1 and Q2 are generally of equal dimensions, then Q1 must be biased under similar conditions to Q2 (i.e. V_{GS1} should be close to zero), and consequently the output voltage from the circuit will be approximately equal to the input voltage.

The small-signal equivalent circuit for the source–follower is shown in Figure 4.8. If Q1 and Q2 are not equal sizes, then the output voltage can be expressed as:

$$v_{out} = \frac{g_m(v_{inp} - v_{out})W_1}{(g_o W_1 + g_o W_2)}$$

The small-signal voltage gain of this circuit then evaluates to:

$$A_v = \frac{(g_m/g_o)}{(g_m/g_o) + (1 + W_2/W_1)}$$

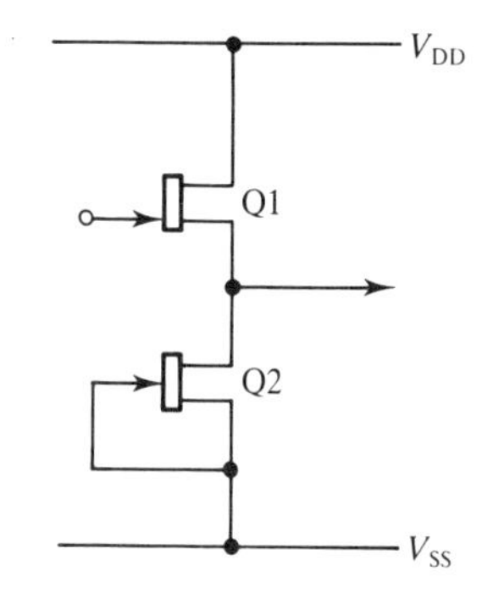

Figure 4.7 A source–follower

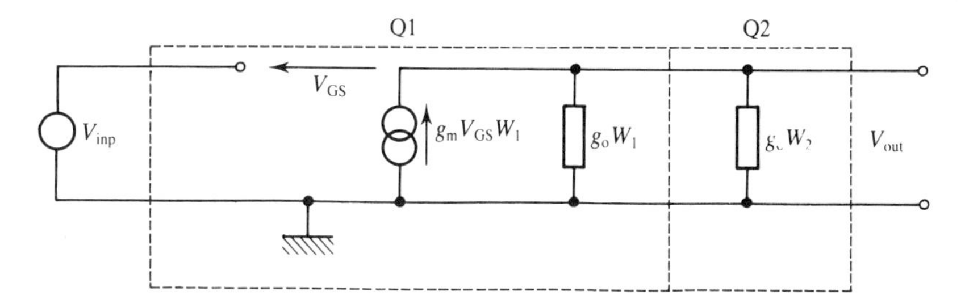

Figure 4.8 The small-signal equivalent circuit for a source–follower

where W_1 and W_2 are the widths of devices Q2 and Q1, respectively, and g_m and g_o apply to unit-width devices. If W_1 and W_2 are equal this becomes:

$$A_v = \frac{(g_m/g_o)}{(g_m/g_o + 2)}$$

Typical values of A_v will then be in the range 0.88 to 0.92.

The voltage gain (or transfer efficiency) is always less than unity so that an a.c. voltage offset ($V_{IN} - V_{OUT}$) appears across the gate–source junction of Q1. If the gain is much less than unity, or the input signal is large, this voltage may forward-bias the gate–source diode, so reducing the input impedance considerably and compromising the ability of the circuit to isolate a high impedance node from a low impedance load. If the input is not forward-biased, then the buffer will exhibit an input impedance equal to the load impedance divided by $(1 - A_v)$; in some circumstances this may mean that two or more buffer stages may be required to prevent the driving circuitry being unduly loaded (see Section 3.12 on output buffers).

The usual applications for this buffer stage will require zero d.c. offset between the input and output. Any offset will reduce the signal handling capability of the circuit. Ideally, assuming devices with infinite output conductance are available, this implies that devices Q1 and Q2 must be of equal width. This arises since the gate of Q2 is connected to its source (i.e. zero V_{GS}) and all the current flowing in Q2 also passes through Q1, hence the gate–source bias (V_{GS}) for Q1 must then set itself to zero. In practice GaAs MESFETs do not have infinite output conductance and thus the drain current is dependent upon the drain–source voltage. The gate–source voltage of Q1 will then only be zero when the drain–source voltages for both devices are equal, i.e. when the input and output voltages are mid-rail. For input values above or below this position, then the gate–source voltage of Q1 will be set slightly positive or negative, respectively, and a non-zero offset will exist. This offset will be superimposed upon the a.c. offset arising from the lower-than-unity voltage gain, and exacerbate any adverse effects. If zero offset voltage is required at some input voltage other than mid-rail, then the ratio of the widths of Q1 to Q2 must be changed to compensate for the non-equal drain–source voltages; for zero offset at input levels positive of mid-rail the width of Q1 should slightly exceed that of Q2, and vice versa.

The input offset voltage will be increased by any mismatch between the MESFETs caused by V_T variations and processing tolerances. Further increase will occur if the load driven by the source–follower is highly capacitive and the input slew rate is high. Device Q1 must supply both the charging current to the load capacitance and the current demanded by device Q2; Q1 must thus adopt a more positive gate–source voltage when charging. Conversely, when discharging the sum of the current through Q1 and the discharge current for the load capacitance is limited by Q2; for any discharge current to flow, Q1 must thus adopt a more negative gate–source voltage. With high capacitance loads, or high input slew rates, the magnitude of the gate–source bias on Q1 may be sufficient to forward-bias the input (and reduce the input impedance) on positive-going transients, and to completely turn off Q1 on negative-going transients. Both effects will introduce serious dynamically varying distortions into the signal. The input offset voltage is a function of the device widths, so the designer must ensure that sufficient widths are chosen so that any capacitance load can be driven without incurring these extreme effects.

4.2.3 Level shifting

Because the threshold voltage of the MESFET is typically non-zero, and usually negative in value, it is often necessary to adjust the d.c. levels of signals in an analog circuit to ensure that the biasing of all parts of the circuit is correct. This adjustment can be simply achieved by a modification of the source–follower buffer described above (see Figure 4.9).

To shift the d.c. level of a signal in the negative direction, an appropriate number of forward-biased Schottky diodes can be inserted between the input and output ports of a source–follower circuit (see Figure 4.9(a)). Although it is possible to vary the voltage drop across each diode by choosing an appropriate current density (determined by the diode and MESFET sizes), the range of adjustment available is rather small. A high a.c. resistance in the forward-biased diode will degrade the transfer efficiency and raise the stage output impedance, and in practice this consideration determines the diode area and perimeter length rather than any diode voltage drop requirement (see Chapter 6). Level shifting obtained by this circuit is thus of a discrete nature, the step size being approximately 0.7 V per diode.

To shift the signal level in a positive direction, the diode chain can be used 'in reverse' (Figure 4.9(b)). The source of Q4 and the drain of Q3 maintain a potential difference

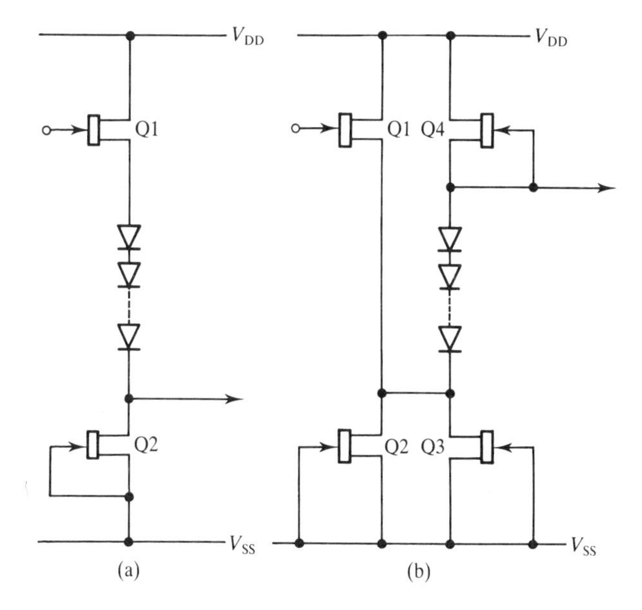

Figure 4.9 Level-shifting circuits: (a) negative level shifting; (b) positive level shifting

determined by the number of diodes in the chain. This circuit has an input impedance which is too low to be driven by most circuits, so the source–follower formed by Q1 and Q2 is used to raise the input impedance. In some circuits, devices Q2 and Q3 may be merged into one (wider) device. Again, discrete steps of approximately 0.7 V per diode must be accepted.

In both the positive and negative level-shifting circuits, the small-signal gain approaches a limit of:

$$A_v = \frac{(g_m/g_o)}{(g_m/g_o) + (1 + W_2/W_1)}$$

when the a.c. resistance of the diodes is negligible. This limit, of course, is the voltage gain of the source–follower circuit alone. Again, the usual configuration has $W_1 = W_2$. The same considerations concerning a.c. and d.c. voltage offsets apply as in the basic source–follower circuit.

4.2.4 Differential amplifiers

Differential amplifiers are often used in preference to the single-ended inverting amplifier described above since the output is dependent upon the voltage difference between the signals to the two inputs, rather than the absolute signal levels. In a differential amplifier, signals which are common to both inputs (i.e. 'common-mode') will not be amplified to the same extent as the signal difference. (The ratio of the differential amplification factor to the common-mode amplification factor is referred to as the common-mode rejection ratio, or CMRR.) This property has many applications, for example, a small signal which is superimposed upon a large d.c. offset can be preferentially amplified by applying the signal to one input of a differential amplifier and a signal equal to the offset to the other input (see Figure 4.10(a)). In this case it is likely that a simple inverting amplifier of the sort discussed in Section 4.2.1 would fail to amplify the a.c. signal at all as the offset would cause the output to 'saturate' at either V_{DD} or V_{SS}. Similarly, signals generated in noisy environments may be very difficult to distinguish using the simple inverting amplifier. If the noise signal is applied to both inputs of a differential amplifier (as shown in Figure 4.10(b)), then it is possible for the required signal to be preferentially amplified and processed.

The circuit operation is illustrated in Figure 4.11. The current defined by the current source is steered in variable proportions through the left and right arms according to the resistance of the switch devices. The larger current will flow in the arm with least resistance; however, the absolute value of the resistance is less important than the resistance ratio. If the resistance is controlled by the input signal voltage, then the current in each arm (and hence the output voltage) will be controlled by the input difference rather than by the absolute levels.

In GaAs differential amplifiers the current source is commonly implemented with a D-MESFET with zero gate–source voltage. The switches may be depletion or enhancement mode MESFETs. As with the simple inverting amplifier, there is also a choice between using active (Figure 4.12(a)) or passive (Figure 4.12(b)) loads which must be made

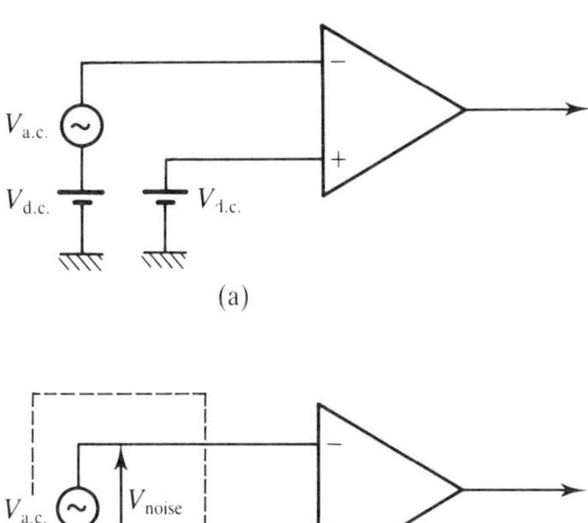

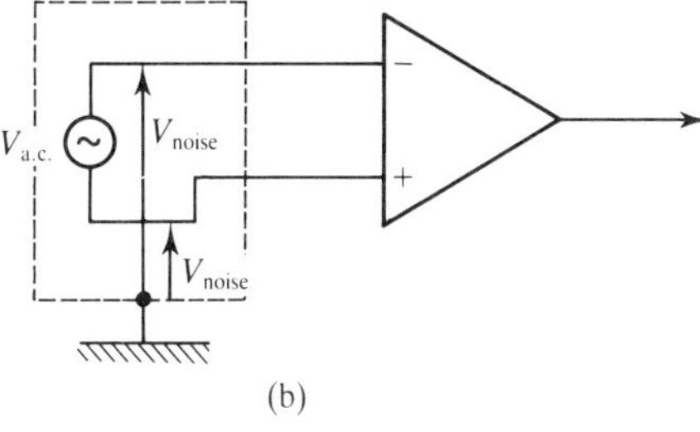

Figure 4.10 Differential amplifier applications: (a) removing d.c. offsets; (b) removing noise

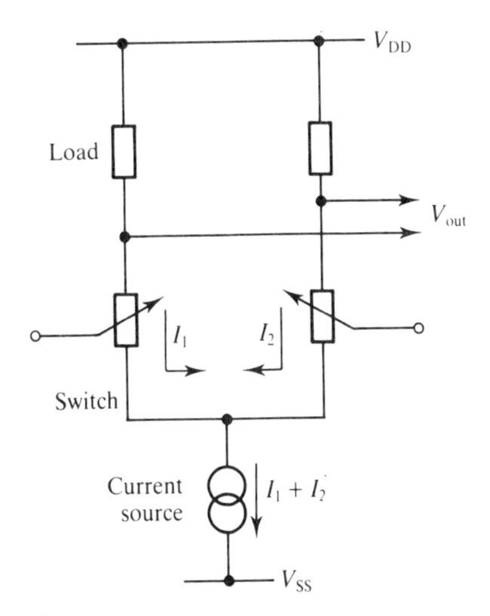

Figure 4.11 Operation of a differential amplifier

depending on the gain and output swing requirements; the active load configuration generally gives a greater gain at the expense of a restricted output swing (the output cannot swing close to V_{DD} since the D-MESFET loads must be kept in saturation if linear characteristics are to be attained).

Care must be taken in choosing the sizes of the devices because of the possibility of forward-biasing the gate electrodes. Referring to Figure 4.12(a), if the current required

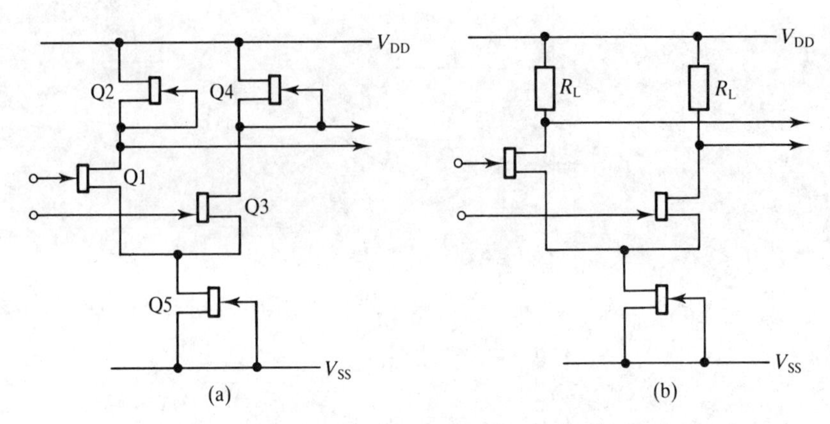

Figure 4.12 Differential amplifiers: (a) active loads; (b) passive loads

by the common current-source device (Q5) ever exceeds the total current available from the load devices (Q2 and Q4), then the difference current will flow in through the gate of one (or both) of the input transistors (Q1 and Q3). This current will obviously reduce the input impedance of the amplifier, and may severely influence the properties of the signal source. As an example of this situation, consider the case where Q5 is wider than Q2 and Q4. If a signal is applied with an amplitude sufficient to turn off Q1 completely, then all the current required by Q5 will have to flow through devices Q3 and Q4. Devices Q5 and Q4 are both connected with a zero gate–source voltage; if equal currents are to flow through both devices then the drain–source voltage must increase across Q4 and decrease across Q5. The decrease of V_{DS} across Q5 will introduce a more positive gate–source voltage on Q3. This is exaggerated by the fact that the magnitude of the V_{DS} increase across Q4 is limited by the input potential (if the drain potential of Q3 falls approximately 0.7 V below the gate (input) potential then it will become clamped by the forward conduction of the gate–drain diode), and so most of the changes enforced by the current continuity requirement will be on the reduction of the drain–source voltage of Q5. With a poor choice of device widths, the drain voltage on Q5 will decrease until Q3 becomes forward-biased, at which point it becomes clamped at approximately 0.7 V below the input potential. The gate current drawn by Q3 will be the difference between the current in Q5 defined at this bias, and the maximum current which can be drawn through Q4.

The common-mode rejection ratio (CMRR) of the differential amplifier is determined by the ratio of the output conductance of the current–source device (Q5) to the load conductances. If it is certain that input signals of an amplitude sufficient to induce gate conduction will never be applied, then a greater width device may be used for the current–source than for the loads in order to obtain an improved CMRR. A trade-off exists between the CMRR and the maximum permissible input swing. Generally, equal device widths are used throughout. Then, in the worst-case condition when (say) Q1 is turned completely off, the currents through Q4 and Q5 can be made equal by changes in the drain–source voltages without incurring gate conduction for all but the most extreme input signal amplitudes.

The circuits shown in Figure 4.12 provide a differential output; with equal size devices the differential gain provided is:

$$A_{v(diff)} = \frac{1}{2}\frac{g_m}{g_o}$$

If only one output connection is used, then half this gain will be lost as each arm effectively only amplifies one half of the applied input differential. For situations where a single-ended output is required, this lost gain can be recovered by using the circuits shown in Figure 4.13. Figure 4.13(a) is an asymmetrical amplifier, while Figure 4.13(b) shows the standard differential to single-ended amplifier arrangement. In the asymmetrical amplifier, devices Q1 and Q4 form a source–follower circuit, and changes on the input of Q1 are thus echoed at the source of Q2 (although with a transfer efficiency of less than unity which is governed by the relative sizes of Q1 and Q4). The gate–source voltage of Q2 thus sees (most of) the full input differential applied to the amplifier; this gate–source voltage is amplified by Q2 and Q3, which form a simple inverting amplifier of the type discussed earlier.

In the standard differential to single-ended amplifier (Figure 4.13(b)), the current through Q5 is divided between the two arms of the circuit by the differential voltage applied to Q1 and Q3. Devices Q3 and Q4 form a simple inverting amplifier so that if (say) a negative signal is applied to Q3 (and positive to Q1), the fraction of Q5 current through the right-hand arm decreases and the output voltage moves towards the positive supply rail. The gate of Q2 is connected to Q4, and thus Q2 has a resistance governed by the output voltage. As the output voltage increases, then the resistance of Q2 falls, and a greater fraction of Q5 current is transferred into the left-hand arm of the circuit thus reinforcing the positive change in the output voltage. In this way the effect of the full differential input voltage is effectively transferred to the right-hand arm inverter formed by Q3 and Q4. The devices Q2, Q1 and Q5 effectively form a source–follower circuit in this amplifier so that the

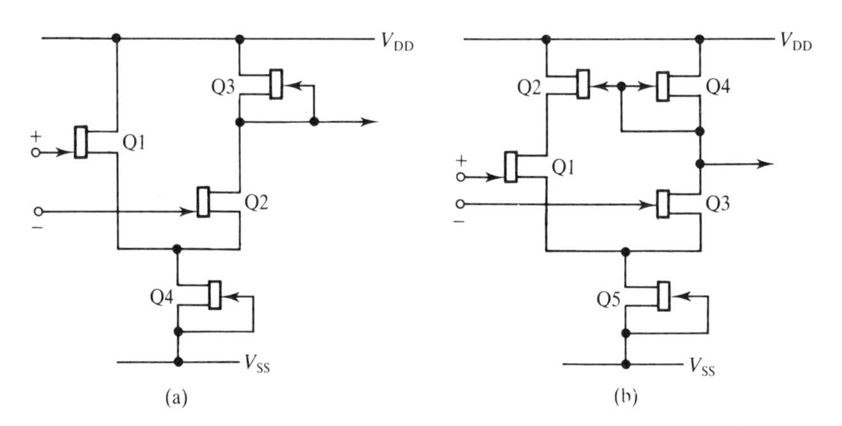

Figure 4.13 Differential to single-ended amplifiers: (a) asymmetrical amplifier; (b) standard arrangement

voltage at the source of Q2 is in phase with and approximately equal to the output voltage at the source of Q4 (the ratio of the voltage at Q2 to that at Q4 is roughly equal to the transfer efficiency of a source–follower, i.e. typically 0.9). The output from this circuit may thus be taken from either arm, although the configuration shown in Figure 4.13(b) gives higher gain and is preferred.

4.2.5 Cascode circuits

The gain of the basic inverting amplifier and the differential amplifiers can be expressed as:

$$A_v = K\left(\frac{g_m}{g_o}\right)$$

where K is a function of the relative sizes of the MESFETs and of the circuit configuration. Generally, K is less than unity, and is typically equal to one-half.

The GaAs MESFET suffers from a relatively low value of g_m/g_o compared to silicon devices, due to surface effects and electron trapping at the channel–substrate interface [38]. Considering the basic inverting amplifier (see Figure 4.5), amplification is obtained by modulating the current flowing through a load resistance. If the input voltage is increased then the current flowing will increase in proportion to the transconductance (g_m) and the output voltage fall. This reduction in the output voltage affects the drain–source voltage of the input MESFET and thus also affects the current flowing, but the action is such as to reduce (or oppose) the increase in current due to the input voltage. The modulation of the amplifier current is thus reduced by the dependence of the current on the MESFET drain–source voltage, i.e. by the output conductance. High gain amplifier stages thus require a high g_m/g_o ratio, which is not generally available from GaAs MESFETs.

One solution to this problem is to use the cascode circuit configuration. In this approach, the circuit is arranged so as to try to maintain a constant drain–source voltage across the current-controlling MESFET. A constant drain–source voltage means that the current modulation due to the input voltage is not reduced by the output conductance. The cascode configuration effectively raises the g_m/g_o ratio for the circuit. Figure 4.14 shows this cascode arrangement applied to the simple inverting amplifier. Provided the supply rail voltages are sufficient to put all the devices into the saturation region (i.e. $V_{DS} > V_T$ for all the devices), then the voltage at the source of Q3 will be close to V_{BIAS} rather than equal to V_{OUT}, as in the simple inverter circuit. The cascode gain is raised from $A_v = (g_m/2g_o)$ to $A_v = (g_m/g_o)$. The width of the cascode transistor Q3 does not significantly affect the gain, but should generally be at least equal to the input (Q1) and load (Q4) widths otherwise its gate–source voltage will be positive; indeed a very small width for Q3 may result in its gate being forward-biased. Generally equally sized devices will be used throughout the cascode circuit to minimize the total area required.

Further gain improvement can be obtained if the drain–source voltage across the load device is also kept constant so as to effectively obtain a very high load impedance. Figure 4.15 shows the cascode circuit modified to achieve this. In this case the amplifier gain is raised to:

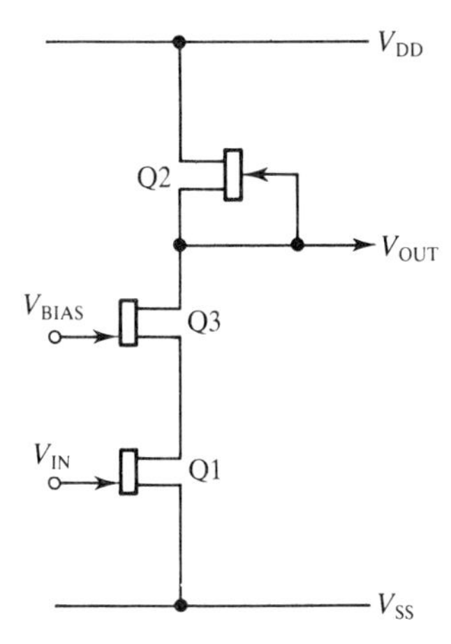

Figure 4.14 An inverting amplifier with cascode input

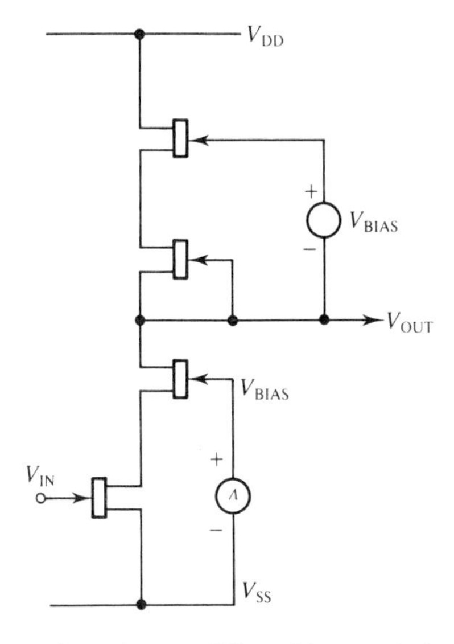

Figure 4.15 An inverting amplifier with cascode input and load

$$A_v = \frac{1}{2}\left(\frac{g_m}{g_o}\right)\left(\frac{g_m}{g_o} + 1\right)$$

Again the supply rail voltages need to be sufficient to put all the devices into the saturation region. The cascode configuration may also be applied to differential circuits and source–follower circuits to considerable advantage if required — this is illustrated in Figure 4.16.

An additional advantage of the cascode stage concerns the input capacitance; when high

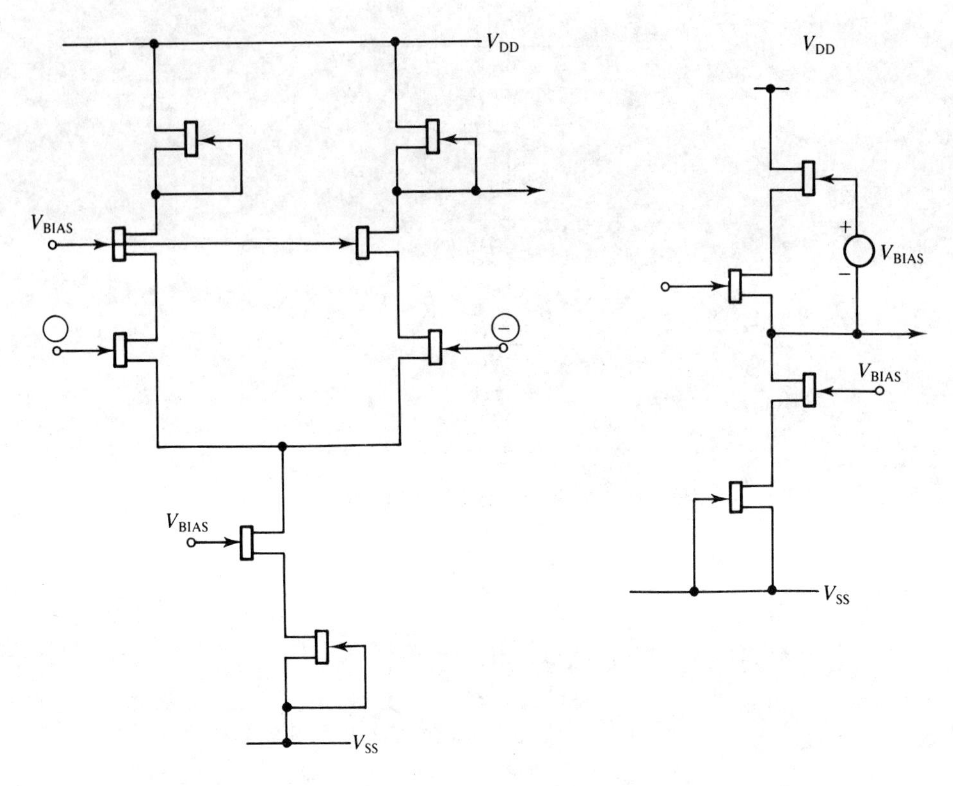

Figure 4.16 Cascode differential amplifier and source–follower circuits

frequency performance is required from an amplifier it is important that this capacitance is minimized as much as possible. In the simple inverting amplifier shown in Figure 4.5, this input capacitance is defined by:

$$C_{INP} = C_{GS} + (1 + A_v)C_{GD}$$

The $(1 + A_v)C_{GD}$ term originates from the fact that any input voltage signal causes a signal which is A_v times larger at the drain. The signal source thus provides a current through C_{GD} which is $(1 + A_v)$ times larger than if C_{GD} were connected between the input and ground; in other words C_{GD} looks like a capacitor of value $(1 + A_v)C_{GD}$ from the input to ground, and the total input capacitance is as defined above. This effective increase in C_{GD} is the well-known Miller effect. The advantage of the cascode circuit is that the drain voltage of the input MESFET remains approximately constant; i.e. it looks like an a.c. ground. There is thus no Miller multiplication of the gate–drain capacitance, and the input capacitance is defined by:

$$C_{INP} = C_{GS} + C_{GD}$$

Obviously the roll-off frequency associated with this input capacitance will be much higher than for the simple inverting amplifier.

4.2.6 Double-level-shift biasing

Biasing of the cascode devices in the circuits described in the previous section may be achieved by several means. For example, an external supply may be used, the bias may be derived from the voltage drop across a chain of diodes, or level-shifting circuits may be used. Figure 4.17 shows the latter two methods used to bias the inverter with cascoded input and load devices.

The major disadvantage of these methods is the large number of components (particularly diodes) and thus the large area required by the bias circuitry. An alternative method known as 'double-level-shift biasing' makes use of the negative gate–source voltage which will be set up in the cascode transistor if it is much wider than the input and load devices; this voltage can be used to bias the switch or load devices as will now be demonstrated.

Consider the circuit shown in Figure 4.18(a). In this circuit the currents through Q1 and Q3 are controlled by the input voltage, and their drain–source voltages are set by cascode devices Q2 and Q4. If Q2 and Q4 are much larger than Q1 and Q3, then their gate–source voltages must be more negative. By cross-coupling the devices as shown in Figure 4.18(b), we can use the negative gate–source voltages on Q2 and Q4 to set a positive drain–source voltage on Q1. The actual magnitude of V_{DS1} will depend on the width ratios W_2/W_1 and W_4/W_3 which determine V_{GS2} and V_{GS4}, respectively. To achieve an

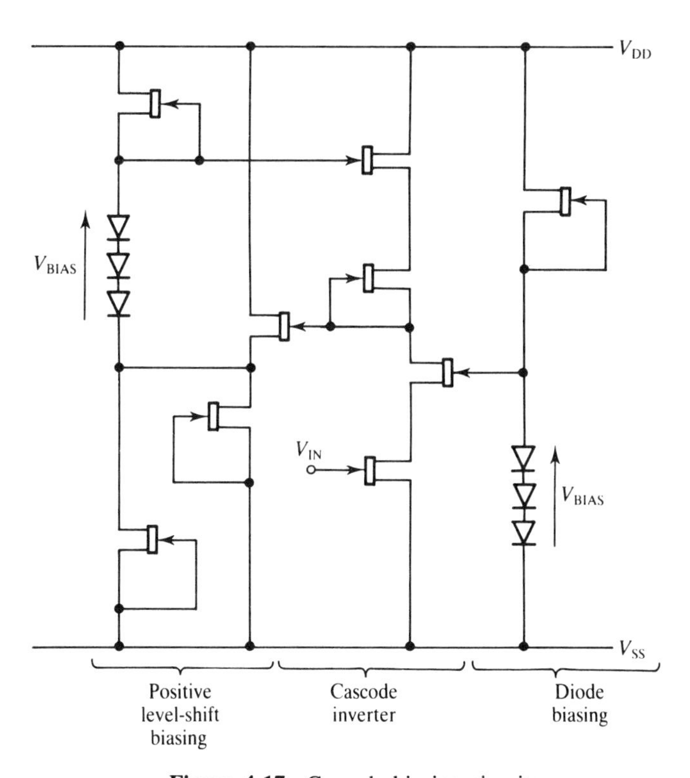

Figure 4.17 Cascode biasing circuits

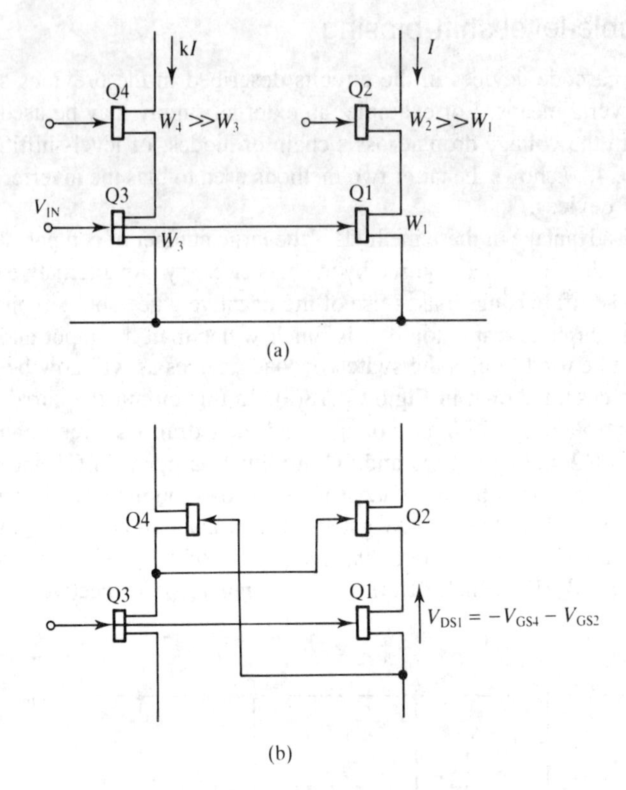

Figure 4.18 A double-level-shift biasing stage: (a) device widths; (b) cross-coupling

increase in gain from this cascode circuit compared to the simple inverter, the drain–source voltage of Q1 must be sufficient to bias it in the saturation region, i.e. $V_{DS1} > (V_{GS1} - V_T)$. If V_{DS1} is to be set by equal contributions from Q2 and Q4, then if $V_{GS1} = 0$ we must ensure that V_{GS2} and V_{GS4} are both greater than $V_T/2$. This can be achieved if W_2/W_1 and W_4/W_3 are greater than 4 (this can be shown using the approximate equations for I_{DS} given in Chapter 2). Alternatively, unequal W_2/W_1 ratios may be used provided the respective gate–source voltages sum to greater than V_T. Since the left arm of the circuit only provides biasing for the right arm, it can use smaller devices than the right arm which must use devices of sufficient width to drive the external load. Reducing the width of the devices in the left arm will give a reduction in the power dissipation. A smaller total width and thus an area advantage can then be obtained if the ratio W_2/W_1 is reduced and the ratio W_4/W_3 is increased. Commonly, values of $W_2/W_1 = 3$, $W_4/W_3 = 10$ and $W_1/W_3 = 4$ are used.

In an inverter, the biasing circuit shown in Figure 4.18(b) may be used directly to bias the input device; however, for the biasing of the load device a slight modification is necessary because the gate voltages of Q1 and Q3 will not be derived from an external

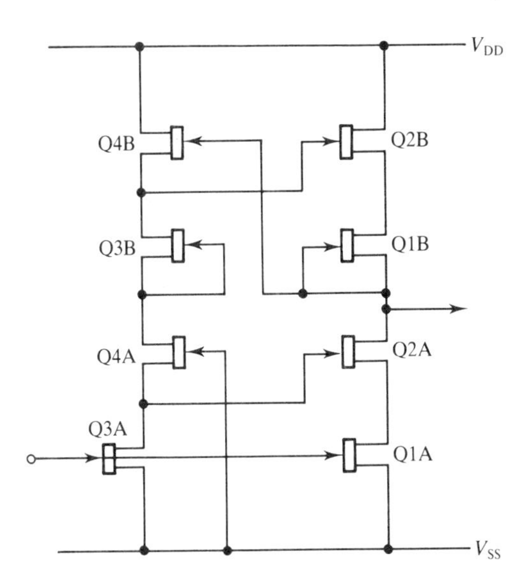

Figure 4.19 Double-level-shift biasing in a cascode inverter

source. Where they are used as load devices Q1 and Q3 will have their gate and source electrodes shorted. The complete circuit for the cascode inverter is thus as shown in Figure 4.19. Both left and right arms in this circuit are identical (barring scaling), hence the voltages at the sources of Q1B and Q3 will track each other and hence the gates of Q1B and Q3B do not need to be interconnected.

A further modification is often made by cross-coupling the gate–source connections of the load devices as shown in Figure 4.20 [22]. This change does not affect the amplifier gain; however, the output rise and fall times may be significantly improved (particularly with large capacitive loads) since now the gate–source voltage of Q1B is not tied to zero volts. If the input signal pulls low then the voltage at the source of Q3B will be able to pull high much quicker than the voltage at the source of Q1B due to its smaller loading. A positive gate–source bias will then be asserted on Q1B, which will consequently be able to supply a greater output current and hence give a faster rise time than the circuit of Figure 4.19. Conversely, a negative gate–source bias will be asserted on Q1B on negative-going output transitions, and faster fall-times will be achieved. This cross-coupled load acts as a push–pull stage similar to that described for digital output buffers in Chapter 3.

Similar biasing techniques can be used in cascode source–follower circuits as shown in Figure 4.21. The devices which control the current flowing in the two arms of this circuit are now in the top half of the circuit rather than the bottom half, but otherwise the circuitry is very similar to that of the cascode inverter. Since the output voltage in this circuit follows the input very closely (more closely than the simple source–follower due to the effective improvement in g_o), this circuit may be modified slightly by deriving

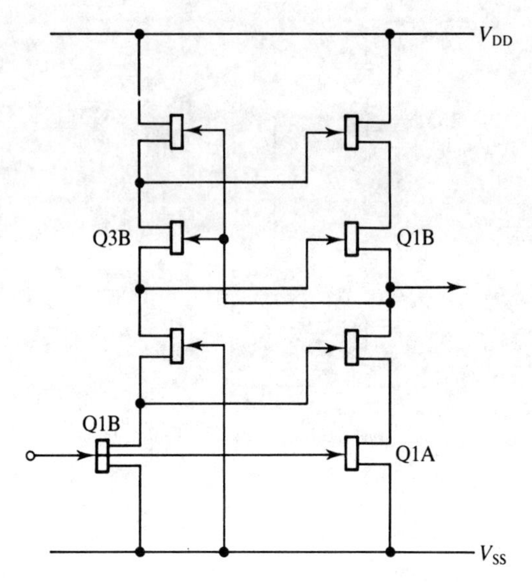

Figure 4.20 A cascode inverter with cross-coupled loads

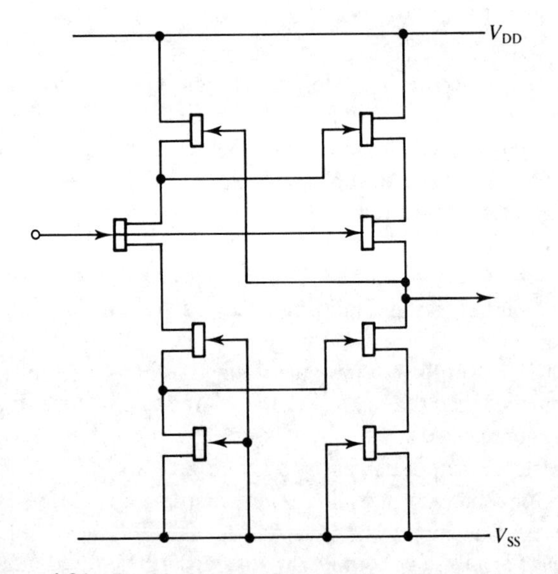

Figure 4.21 Double-level-shift biasing in a source–follower

the control voltage for the current in the left arm from the output rather than directly from the input. This is shown in Figure 4.22. This circuitry bears some similarity with that of the inverter with cross-coupled loads, although the output properties are not significantly affected in this case.

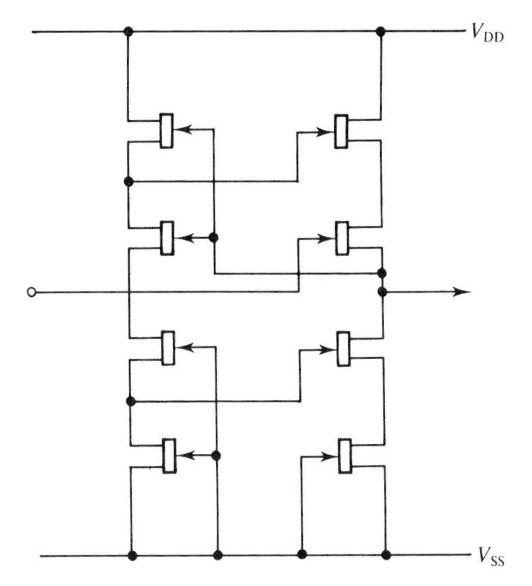

Figure 4.22 A cascode source–follower with modified input connection

4.2.7 Double-cascode circuits

Although the use of a cascode circuit does significantly improve the gain, the inherently low value of g_m/g_o (typically in the range 15–25) may still yield values of gain which are too low for many applications. Further improvements can be obtained by cascoding the cascode devices as well, as shown in Figure 4.23. Using this 'double cascode' technique can result in inverter gains of approximately $A_v = (1/2)(g_m/g_o)^3$.

Biasing of the double-cascode circuit can be achieved using the methods discussed above. Figures 4.24 and 4.25 show double-level-shift biasing applied to a double-cascode inverter (with cross-coupled loads) and a source–follower [22, 39]. The same restrictions on the width ratios apply as in the single-cascode circuits. All the devices in the right-hand output arm must operate in the saturation region and this may necessitate an increase in the supply rail voltages (and hence an increase in the power dissipation) compared to the simple inverter and cascode inverter circuits.

4.2.8 Analog switches and drivers

GaAs MESFETs, like silicon MOSFETs, can be viewed simply as voltage controlled resistors. By switching the channel resistance between a low resistance (closed) state and a high resistance (open) state, the device can be used as an analog switch to control the flow of signals within a circuit. The high electron mobility of GaAs relative to silicon gives lower resistance switches for structures of similar area (and thus capacitance), and hence a higher signal bandwidth. One major difficulty exists, however, regarding the use

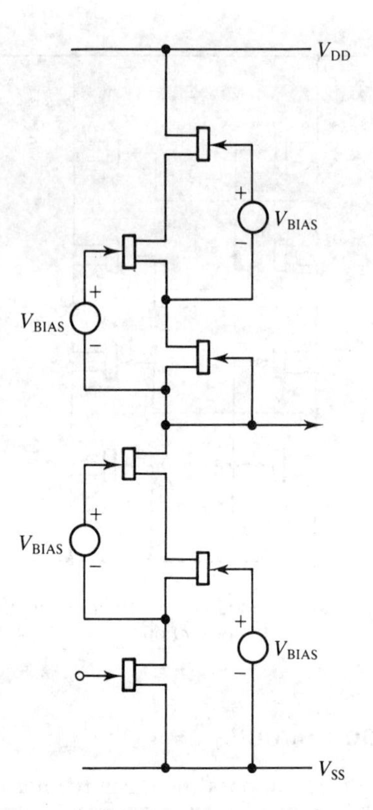

Figure 4.23 A double-cascode inverter

of MESFETs as analog switches, and this arises from the fact that it is possible to get conduction through the gate electrode if it is forward-biased. Hence interaction between the switch control voltage (the 'clock') and the switched analog signal may occur. The level of gate conduction is non-linearly dependent upon the analog signal and the clock levels, and thus the resulting distortion cannot be removed by filtering. The possibility of gate conduction in the switch must be avoided. Generally depletion-mode devices are used for GaAs analog switches since it is difficult to obtain low values of on-resistance in enhancement devices without approaching the forward-biased condition. A specific switch-driver designed to ensure correct operation at all analog signal levels is also often used.

The clock level closing the switch should be as positive as possible to give a low on-resistance, while not allowing the gate–channel bias to approach 0.7 V. Usually a maximum level equal to the analog signal level is used (i.e. a gate–channel bias of zero volts), but the gate–channel bias should generally not exceed +0.3 V. The clock level opening the switch should be sufficiently more negative than the most negative analog signal level to ensure that the gate–channel bias is below the threshold voltage in this situation. If the analog signal level is small (typically less than 200 mV), and has a well-defined d.c. component, then gate conduction can be avoided by using clock levels

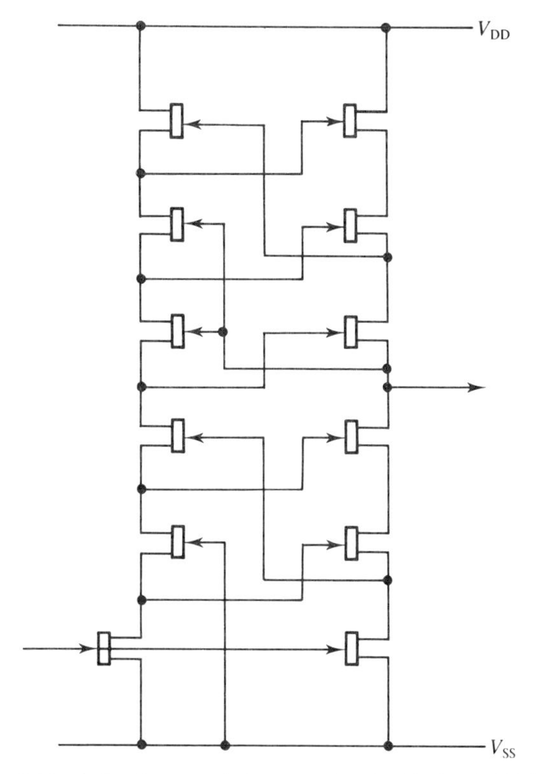

Figure 4.24 A double-cascode inverter with double-level-shift biasing and cross-coupled loads

appropriately regulated to meet these requirements (see Figure 4.26). For large signals, however, this arrangement would either yield high values of on-resistance at the most positive signal excursions (due to a negative gate–channel bias) or cause forward gate conduction at the most negative signal excursions (due to a positive gate–channel bias). A switch-driver circuit is required to ensure that the maximum clock level tracks the analog signal level; the use of an on-chip switch-driver also makes it easier to deliver signals with fast transitions to the switches due to the reduced parasitics associated with on-chip wiring compared to those associated with external switch-drivers (see Chapter 6).

Figure 4.27 shows one possible implementation of a switch-driver. The clock input is provided by an inverting amplifier whose power supply is derived from the analog signal via a source–follower stage. The control signal will thus swing between the output voltage from the source–follower in the closed-state (which will equal the analog signal minus the input offset voltage of the source–follower), and a low (open) level defined by the ratio of the inverting amplifier input and load resistances. The driver for the analog switch must be designed so that this latter level always pulls the analog switch gate–channel bias below the threshold voltage for all possible analog signals (the worst-case is at the most negative signal level). This can be achieved by appropriately ratioing the inverter input and load widths as described in Chapter 3.

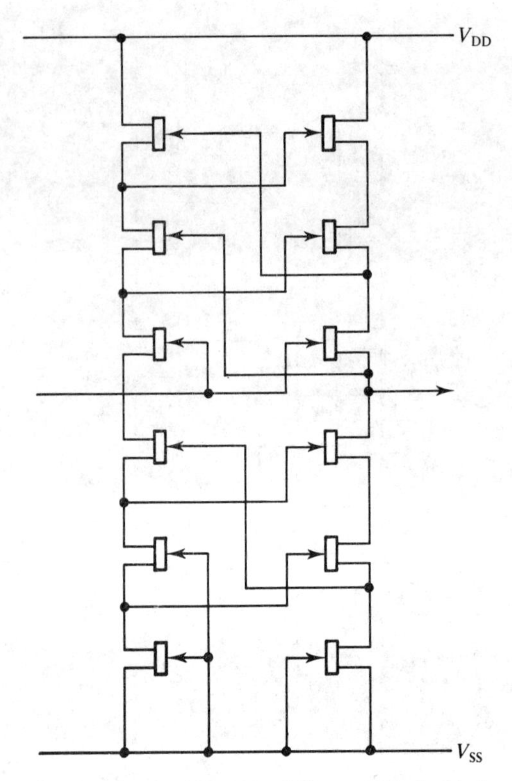

Figure 4.25 Double-cascode source–follower with double-level-shift biasing

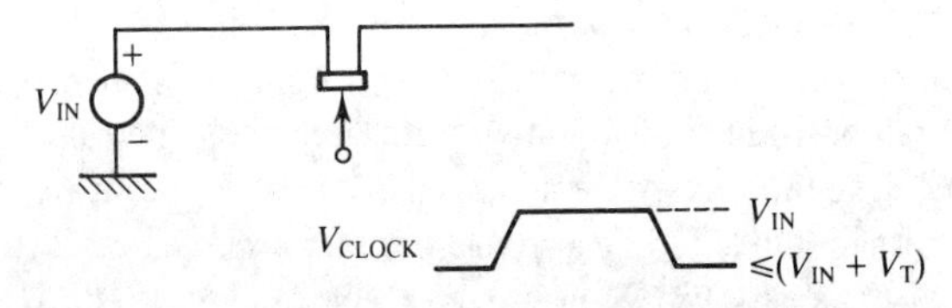

Figure 4.26 Clock signal levels for MESFET analog switches

The source–follower stage must be designed so that its input offset voltage, which is dependent upon the analog signal level (see Section 4.2.2) does not approach plus or minus 0.7 V (and preferably lies within plus or minus 0.3 V) when the analog switch is both open and closed. If the gate–source voltage at the input of the source–follower (the input offset voltage) approaches +0.7 V, then severe loading of the input signal source will occur due to the reduced input impedance in this condition. If the input offset voltage approaches −0.7 V then the gate–channel diode in the switch will be become forward-biased which is undesirable for the reasons described above. The input offset voltage depends on the current flowing through the source–follower, and will be most positive

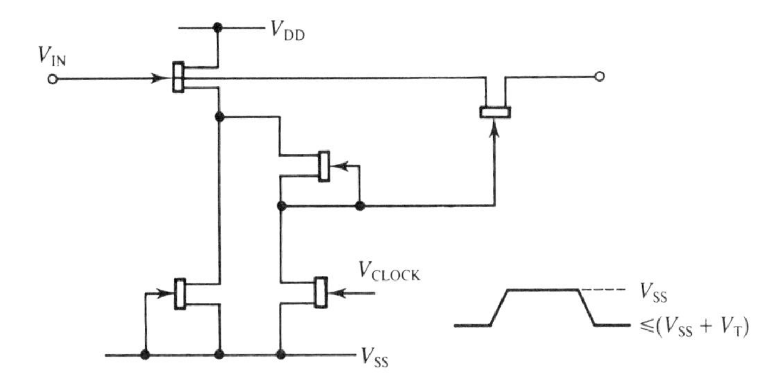

Figure 4.27 A buffered switch-driver for large signal analog switches

when the analog switch is open, as the inverting driver will be drawing current in this situation, and most negative when the analog switch is closed. The negative offset in the closed state can be reduced by appropriately ratioing the widths of the devices in the source–follower as described in Section 4.2.2. The positive offset voltage in the open state can be reduced by scaling the source–follower devices so that the current drawn by the inverter is small in comparison with the source–follower current.

Typically, the source–follower should use device widths at least five times greater than the inverting switch-driver to meet the criterion on input offset voltage limits. However, the level of current creates a trade-off between the offset voltage and the power dissipation. If a circuit has a large number of switches then the power dissipation may have to be limited, which in turn will impose a minimum offset level which must be accepted. A modification which improves on this situation makes use of the SCFL inverter, as shown in Figure 4.28. This SCFL inverter draws a current which is constant in both states of the analog switch, and this inverter replaces the current–source device in the source–follower circuit. The input offset voltage will now be constant in both states, and the offset can be minimized simply by appropriate ratioing of the source–follower device Q1 and the SCFL current–source device Q2. It is not necessary to use very large devices in this circuit, and hence the power dissipation is much reduced, although the SCFL inverter does impose an increased component count and requires a more negative supply rail.

Another alternative switch-driver, which reduces the power dissipation still further, is based upon the removal of the buffer entirely so that the inverter is powered directly from the signal line (see Figure 4.29(a)). This unbuffered circuit will load the signal when its input is high; however, since the switch will be open in this state the loaded signal will be isolated from the subsequent circuitry. A simple implementation of this circuit is shown in Figure 4.29(b).

Although the buffered and unbuffered switch-drivers discussed above remove the potential of forward-biasing the gate–channel diode in the switch, these circuits still have an undesirable characteristic in the form of a level of clock feedthrough which is dependent upon the analog signal level. The clock feedthrough is charge that is introduced into the signal circuit by the changing clock signal through the parasitic capacitance associated with the gate of the switch device (see Figure 4.30(a)). The amount of charge introduced

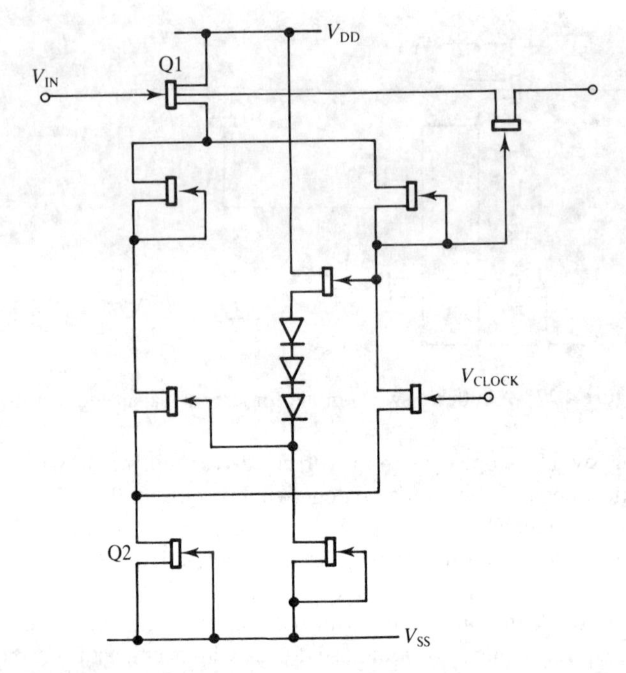

Figure 4.28 A buffered SCFL switch-driver for analog switches

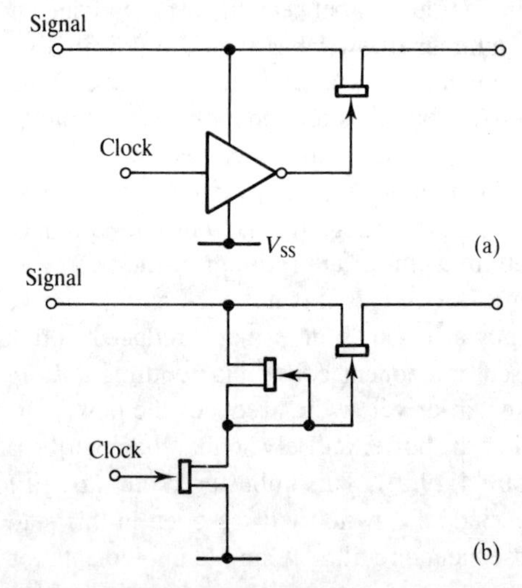

Figure 4.29 An unbuffered switch-driver: (a) block diagram; (b) simple inverter configuration

is proportional to the clock signal swing, which in turn depends upon the analog signal level. An unbuffered driver will also introduce charge in proportion to the change in the signal level, which occurs due to the driver loading (see Figure 4.30(b)). Since the loading effect depends upon the current drawn by the driver, which in turn is dependent upon the signal level, this additional source of feedthrough is also signal-dependent.

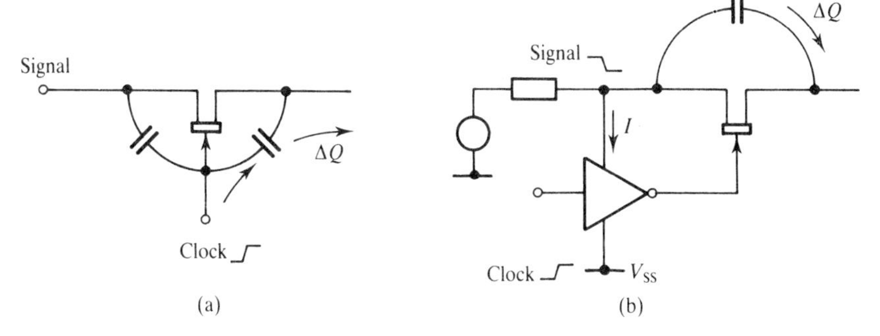

Figure 4.30 Clock feedthrough due to (a) parasitic capacitance, and (b) signal loading

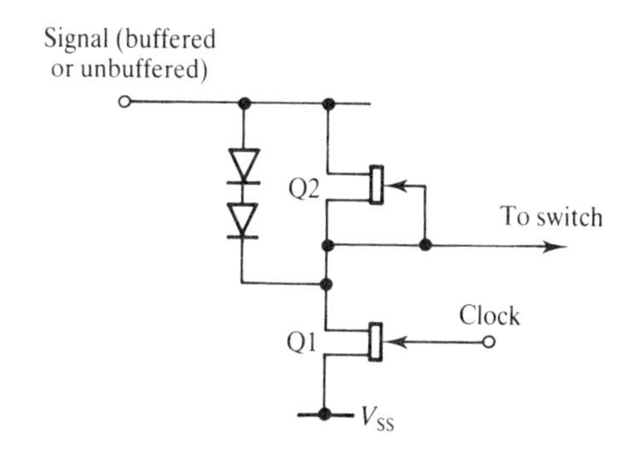

Figure 4.31 A switch-driver with diode-limited signal swing

The situation can be improved by modifying the driver through the introduction of a diode chain in parallel with the driver load device (Figure 4.31) [40]. The voltage swing available at the output of this driver is now constrained by the forward voltage drop across the diode chain. This swing is almost independent of the signal level, barring a variation which occurs due to the fact that the diode drop is current-dependent, and the current is signal-dependent according to the output conductance of the inverter input device Q1. A further modification, introducing a cascode device as shown in Figure 4.32, makes the driver current practically signal-independent, and thus a constant output signal swing is achieved along with a constant load on the analog signal source in the case of the unbuffered driver. This cascode switch-driver [41] thus produces an almost constant level of clock feedthrough which can be easily filtered from the analog signal.

4.3 Design examples

4.3.1 An operational amplifier

Figure 4.33 illustrates the possible design of a wide bandwidth operational amplifier

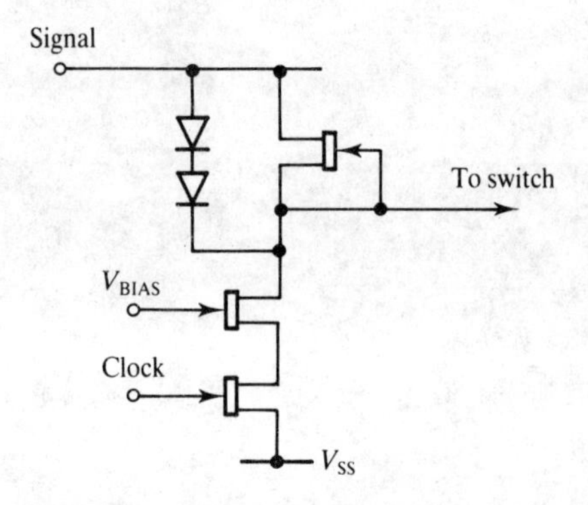

Figure 4.32 A cascode switch-driver circuit

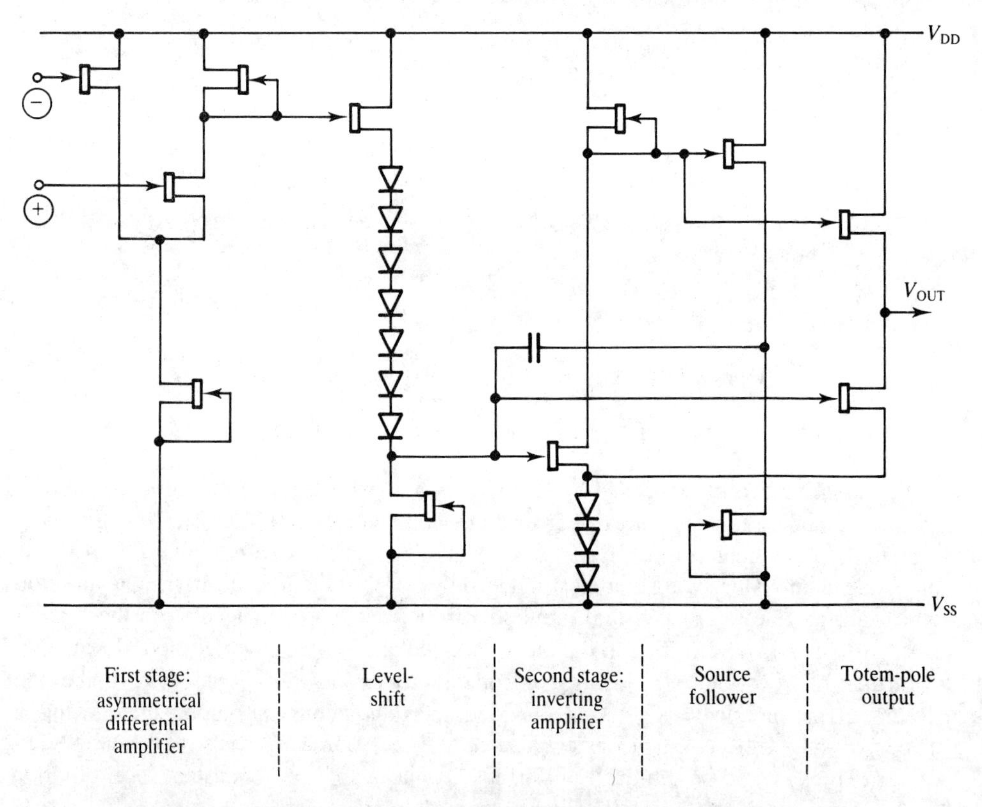

Figure 4.33 An operational amplifier design ($V_{DD} = +5V$, $V_{SS} = -5V$)

designed using GaAs MESFETs and the building blocks discussed in the previous sections [42].

The first stage of this amplifier is an asymmetric differential-to-single-ended amplifier stage driving a second stage simple inverting amplifier to give a good overall gain. The two stages are connected by a negative level-shifting stage; this stage also buffers the high output impedance of the first stage from the input capacitance of the second stage. The negative rail for the second stage is arranged to be approximately two volts above V_{SS} by biasing through a chain of three, large area (i.e. low resistance) diodes. This ensures that the current–source MESFET in the level shift network is kept in saturation. To allow maximum dynamic range, the quiescent output level of the first stage should be roughly mid-way between the positive supply rail and the voltage at the top of the current–source device (which will be close to the amplifier d.c. input level). Similarly, the output level of the second stage should be mid-way between its positive and negative supply rails, and with equally sized devices this will require the input gate–source voltage of the second stage to be close to zero; i.e. the input level will be approximately 2 V above V_{SS}. The level-shift stage must allow coupling between the output of the first stage and the input of the second, and thus the number of diodes in this stage is defined by the need to allow this coupling.

In any two-stage amplifier design it is possible for the phase-shift between the inverting input and the output to reach zero degrees at high frequencies while the gain is greater than unity. Any feedback from the output to this input (which would normally be negative feedback) will then be in phase and could cause the circuit to oscillate unless appropriate remedial action is taken. The remedy in this circuit is provided by a 'compensation capacitor' connected around the second stage; this capacitor introduces local negative feedback into the second stage at high frequencies, thus ensuring that the overall amplifier gain is reduced to unity well before the phase-shift approaches zero degrees. Undesirable feedforward effects (whereby signal bypasses the second stage via the compensation capacitor) are reduced by driving the capacitor via a source–follower stage; this source–follower allows feedback signals to pass from the second-stage output to the input, but effectively shorts all feedforward signals from the second-stage input down to the negative supply rail.

The output from the amplifier is taken from a 'totem pole' push–pull stage, similar to that discussed for DCFL output drivers in Chapter 3. In this case depletion-mode MESFETs are used throughout, and to turn off the lower D-MESFET its gate–source voltage must be taken below its negative threshold voltage. If the source of this MESFET were connected to the V_{SS} rail, as in the DCFL output driver, this would require the gate bias to be taken below V_{SS}, i.e. another (negative) supply rail would be required. However, in this circuit this is avoided by arranging that the source voltage sits approximately 2 V above V_{SS} by connecting to the same chain of diodes used to bias the source of the second-stage inverter.

The gain of this amplifier is determined by the product of the gain of the first and second stages, and the bandwidth by the roll-off imposed by the compensation capacitor. Without this compensation, the intrinsic amplifier phase shift and bandwidth are determined primarily by the input (Miller) capacitances in the first and second stages. If higher gain is required then the cascode techniques described previously can be employed. This will have the additional advantage of reducing the Miller effect on the input capacitances, and thus reducing the phase shift and raising the intrinsic amplifier bandwidth. If compensation is required to ensure stability when feedback is applied, a much smaller compensation capacitor may be used than in the simple non-cascode amplifier, and the bandwidth of the compensated amplifier may also be much improved [43].

4.3.2 A switched-capacitor filter

Besides digital and analog applications, a further area where GaAs offers a significant advantage over silicon is in sampled-analog applications such as switched-capacitor filters (SCFs) [44]. These circuits allow high accuracy filters to be integrated rather than being built from discrete components, which gives all the advantages usually associated with ICs (e.g. cost, reliability, size, design security, etc.). The theory of SCF design is beyond the scope of this book, suffice it to say that for high speed and good accuracy then high gain fast-settling amplifiers and fast-switching low power switches are required. Figure 4.34 shows the architecture of an SCF with a bandpass frequency response [45]; this filter is designed to be driven by a two-phase non-overlapping clock, with a clock to centre frequency ratio of 25 : 1. The pass-band attenuation should be 0 dB, and the Q-factor should be 16. This specification can be met using the double-cascode inverting amplifier (see Figure 4.24) and a mixture of simple inverter and unbuffered cascode switch-drivers (see Figures 4.29 and 4.32). The cascode switch-driver is used in those situations where the analog switch has a high dynamic range, such as at amplifier outputs and the filter input. At the amplifier inputs the signal level is effectively constant (at a 'virtual V_{SS}' level) and the simple inverter switch-driver is used. The use of the cascode driver at these latter points would necessitate an extra, more negative, supply rail in order to ensure all the cascode driver devices operate in the saturation region, which would increase the power

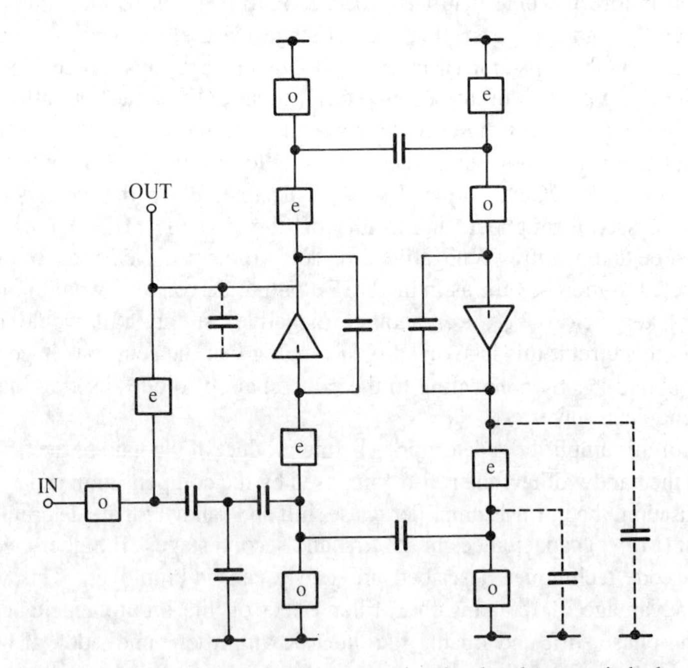

Figure 4.34 Architecture of a second-order bandpass switched-capacitor filter

dissipation. The load and the input device in the simple inverter driver are scaled so as to provide a signal swing equal to that delivered by the cascode driver (which is set by the diode chain), so that the components of clock feedthrough from opposite sides of the switched-capacitors cancel.

Close examination of the SCF will show that the amplifier capacitive load changes according to the state of the clock signals. The settling time of the amplifier exhibits a minimum, as shown in Figure 4.35, with an unstable region at very low loads. Each amplifier is thus scaled so as to optimize the settling times in the critical clock states, and 'dummy' capacitors are added to ground to ensure stability in other non-critical states (particularly in the non-overlapping guard-band period when all the switches are open); these dummy capacitors are shown with dashed connections in Figure 4.34. The filter architecture is 'parasitic-insensitive' so that these extra capacitors do not affect the filter frequency response.

Since the capacitive load is critical in determining the settling time, it is essential that the filter is isolated from the effects of any load connected to the output. This is provided by a double-cascode source–follower (see Figure 4.24). Two of these stages connected in series provide extremely good buffering, and the insertion of a cascode switch-driver and capacitor allows sample-and-hold operation to be provided if required.

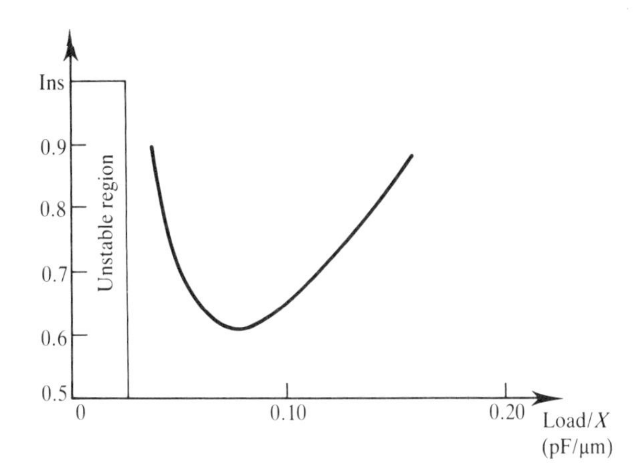

Figure 4.35 Settling time of the double-cascode inverting amplifier (width of load and input MESFETs = X μm)

5 □ Processing

The manufacture of GaAs ICs bears a great deal in common with silicon IC manufacturing. However, there are many essential differences. This chapter describes the manufacturing process of GaAs ICs with emphasis on these differences.

5.1 Crystal growth

The first requirement in the manufacture of a GaAs IC is for GaAs wafers. In common with silicon IC processing, this requires the growth of large crystals which can be cut into wafers. These wafers are required to have the following properties:

(a) a low dislocation (defect) density since crystal dislocations will affect the device characteristics and thereby limit the yield (see Section 5.1.2);
(b) large round wafers to increase the number of chips per wafer;
(c) good flatness to aid the resolution of extremely small features by photolithography (see Section 5.4);
(d) cleanliness to limit the introduction by dust particles of device defects in the fabrication process;
(e) a high undoped resistivity (greater than $10^7\ \Omega$ cm) to give good device isolation.

The growth of GaAs crystals is complicated by the fact that while growth temperatures in the range 300–600°C are commonly used, appreciable dissociation of arsenic can occur even at temperatures below this. Some means of suppressing the loss of arsenic from the crystal must be achieved as gallium-rich GaAs will have more crystal dislocations and a lower resistivity than stoichiometric (50 : 50) GaAs. There are currently two major techniques used for the growth of GaAs crystals: the horizontal Bridgman technique (HB) and the liquid encapsulated Czochralski method (LEC). These techniques are described below.

5.1.1 Horizontal Bridgman (HB) crystal growth

The secret to the successful growth of HB grown crystals is to suppress the loss of arsenic from the crystal by initiating crystal growth in an arsenic atmosphere. The growth process can be broken down into three stages which are illustrated in Figure 5.1. In the first stage gallium and arsenic are sealed in an evacuated quartz chamber which is placed inside a controllable heater. A GaAs seed crystal is also sealed in the chamber at the edge of the boat containing the gallium.

The second stage involves raising the temperature of the chamber so that the arsenic source and the GaAs seed crystal are below the melting point of GaAs, and the gallium boat is above this melting point. Arsenic is highly volatile so that the chamber becomes pressurized with arsenic gas at this high temperature; reaction of the gallium and arsenic occurs in an arsenic atmosphere to form molten GaAs.

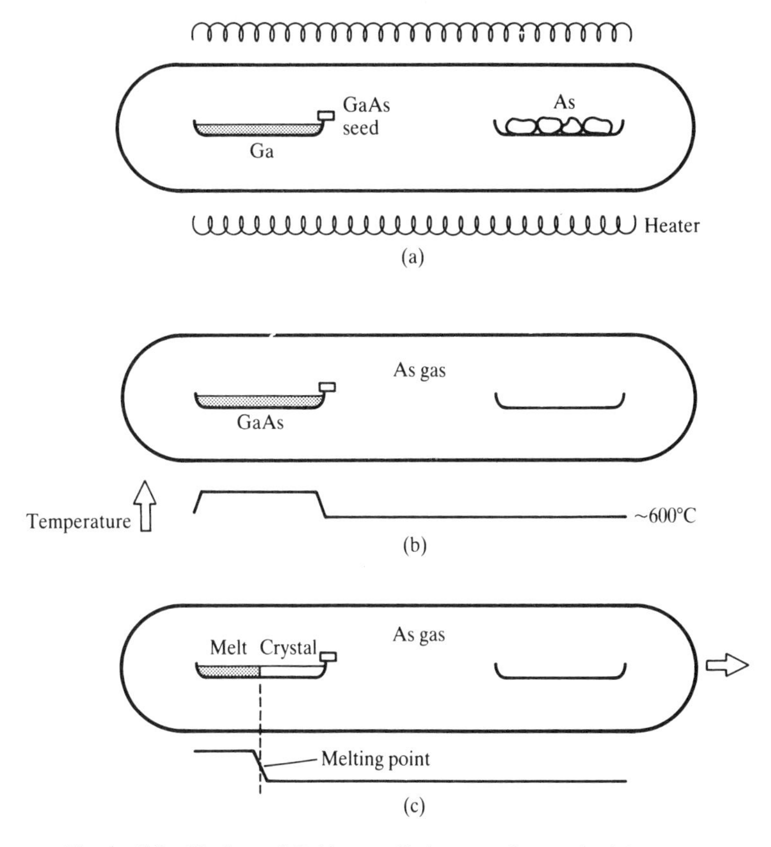

Figure 5.1 Horizontal Bridgman GaAs crystal growth: (a) vacuum sealing; (b) synthesis; (c) crystal growth

The final stage is the growth of a GaAs crystal which is achieved by slowly drawing the chamber out of the heater. As the temperature of the melt is lowered below the melting point crystallization will occur; the seed crystal ensures that the resulting growth is of the required orientation.

The resulting HB-grown crystal often suffers from silicon impurities (which act as *n*-type dopants) from the quartz boat used to contain the melt. These impurities are generally counteracted by doping the melt with chromium [46], although this may cause problems as the chromium can redistribute in subsequent processing giving unpredictable behaviour. Some success has been achieved using a pyrolytic boron nitride (PBN) boat or a hydrogen atmosphere to achieve undoped or low doped material [47].

The cross-section of the wafers is D-shaped; generally subsequent processing equipment prefers circular wafers which means that some trimming of the wafers has to be carried out and thus some material is wasted. The size of wafer which can be grown is also limited to less than 3 in. at present. This should be compared with silicon technology which commonly uses 4–5 in. wafers and is moving towards 6 or even 8 in. material.

On the plus side, HB crystals generally have a relatively low dislocation density (less than $10^3\ cm^{-2}$), and the equipment and subsequent wafer costs are relatively low. Generally HB-grown crystals are used for those applications where the effects of Cr-doping are not critical, for example, opto-electronic or microwave applications where the active devices are commonly fabricated in an epitaxial layer grown on top of the substrate.

5.1.2 Liquid encapsulated Czochralski (LEC) crystal growth

LEC-grown crystals rely upon the use of a layer of molten boric acid (B_2O_3) on top of the GaAs melt to suppress the loss of arsenic. The melt is formed by reacting gallium and arsenic in a slowly rotating crucible as shown in Figure 5.2. The rotation helps to reduce any thermal gradients parallel to the melt surface which might be caused by heater non-uniformities. A similarly rotating seed crystal of GaAs is then lowered into the melt and then slowly pulled out, pulling with it a crystal from the melt. The resulting crystal area is dependent on the rate of pulling.

This method of GaAs crystal growth has the advantage of yielding large round wafers; 3 in. wafers are now common and trials on 5 in. wafers have been made (although presently yielding very high dislocation densities). The dislocation density is generally higher than in HB-grown crystals; levels of $10^4\ cm^{-3}$ or greater are typical. Substantial improvements can be achieved by reducing the thermal gradient at the solid–liquid interface and by monitoring the water content in the boric acid layer: 3 in. wafers with dislocation densities in the range 0–200 cm^{-3} have been achieved [48]. These wafers have an improved electron mobility and better threshold uniformity, both factors are essential if fast LSI GaAs circuits are to be produced.

As with HB-grown crystals, LEC-grown crystals may also suffer from silicon impurities if a quartz crucible is used. The use of PBN can again reduce this contamination so that high resistivity undoped crystals can be produced. Although there is no clear choice between HB and LEC-grown crystals, LEC appears likely to become the dominant process due

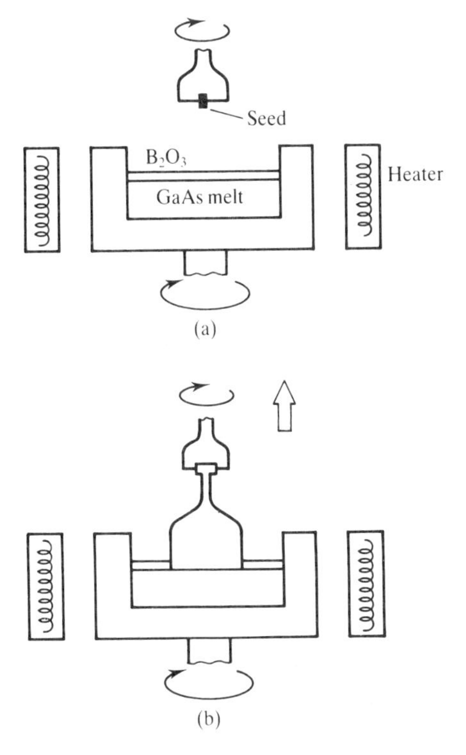

Figure 5.2 Liquid encapsulated Czochralski (LEC) GaAs crystal growth: (a) synthesis; (b) crystal growth

to its ability to produce larger round wafers, provided the production of low dislocation densities can be maintained.

The availability of low dislocation densities in wafers used for GaAs IC production is important since it has been shown that high dislocation densities produce high variations in the threshold voltage (V_T) of the MESFETs [49], which thus affects the yield available from LSI/VLSI circuits (see Chapter 3). Attempts to correlate the MESFET threshold voltage with the distance to the nearest dislocation defect have achieved variable results [50, 51], and it is now generally accepted that many other factors (such as the surface treatment [52]) also contribute to the variability in device characteristics, so that variations in V_T can still be expected in dislocation-free ('zero-defect') wafers. However, low dislocation densities are an important pre-requisite for minimizing the device variations, and much of the effort in the field of GaAs ICs is directed towards improving material properties and IC processing, and not just towards circuit designs techniques.

Alternative methods which have been used to reduce the dislocation density include:

(a) applying a strong d.c. magnetic field to the melt to suppress convection currents and thus stabilize the solid–liquid interface [53], and
(b) adding In or InAs (approximately 1%) to the melt [54].

The latter technique looks extremely promising, and 3 in. crystals have been produced having large defect-free areas [55]. These wafers show an improvement in the uniformity of V_T, although problems do exist in trying to maintain uniform levels of In-doping throughout the crystal and in preventing the onset of polycrystallinity during the growth process.

5.2 Active layer preparation

Once good, high resistivity GaAs wafers have been prepared, the next step is to prepare a doped region in which the devices will be defined. This generally involves either:

(a) the growth of an epitaxial layer by vapour-phase epitaxy (VPE), metal–organic chemical vapour deposition (MOCVD) or molecular beam epitaxy (MBE) techniques, or
(b) ion implantation.

These techniques are described below.

5.2.1 Vapour-phase epitaxy (VPE)

Without going into excessive detail, the process of growing GaAs epitaxial layers by VPE is generally either by a chloride [56] process or a hydride [57] process. These processes are shown schematically in Figure 5.3. In both processes GaCl is reacted with As gas

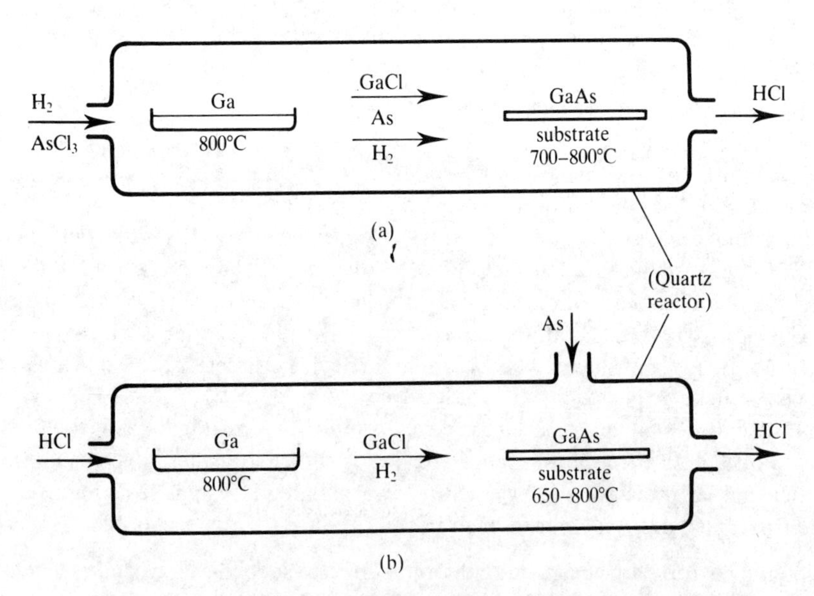

Figure 5.3 Vapour-phase epitaxy (VPE): (a) chloride VPE; (b) hydride VPE

and hydrogen to form GaAs (which is deposited onto a GaAs substrate) and HCl. The differences involve the way the GaCl is formed and the point at which As is introduced. Generally the hydride process is easier to control and is thus preferred.

Dopants are introduced as gaseous sources of $SiCl_4$, H_2S or H_2Se for *n*-type material, or Zn or $ZnCl_2$ for *p*-type GaAs. An undoped buffer layer may often be grown between the substrate and the conducting active layer to reduce the influence of any defects in the substrate. In this case, the quality of the substrate becomes much less important.

5.2.2 Metal–organic chemical vapour deposition (MOCVD) [58]

The apparatus used for MOCVD growth is similar to that used for VPE; however, the key differences are the reactants used and the method of substrate heating. The reactants are a gaseous metal alkyl (e.g. trimethyl gallium) and arsine hydride, while the substrate is heated by a radio-frequency (rf) coupled substrate holder; the chamber walls are also at ambient temperature in contrast to VPE where the chamber is at the same (or higher) temperature as the substrate.

Use of a gaseous source of gallium allows very good control of layer thickness to be achieved, and the low temperature of the chamber walls allows the growth of aluminium compounds which would otherwise react with the quartz; for example, it is possible to grow $Al_{(1-x)}Ga_xAs$ multi-layers on top of GaAs layers, a process which is essential for the production of lasers and other advanced devices (see Chapter 7).

MOCVD is not yet a high volume production process, but does show promise of development.

5.2.3 Molecular beam epitaxy (MBE) [59]

In MBE layer growth, molecular or atomic species are obtained from heated sources (Knudsen effusion cells) and are allowed to impinge upon a heated substrate in an ultra-high vacuum. Shutters in front of the sources allow control of the type of molecular beam, and control of the substrate and source temperatures allows good control of the grown layer stoichiometry, doping level and composition. To minimize contamination from impurities the substrates are introduced via an 'airlock' so that the vacuum can be maintained between growth runs. A schematic of the apparatus is shown in Figure 5.4.

It is possible to grow extremely complex devices by MBE, with complex doping profiles and heterojunctions (e.g. AlGaAs/GaAs) if desired. Control of the growth process is generally better than in the MOCVD process. The major drawback to this process is the long cycle times and thus the unsuitability at present for high volume IC production.

5.2.4 Ion implantation

Ion implantation involves ionizing a required source (generally by heating), purifying the produced ions by passing them through what is effectively a mass spectrometer, and then accelerating the selected species towards the substrate to be implanted. The purification stage is necessary to remove any unwanted isotopes from the ion stream, since the doping

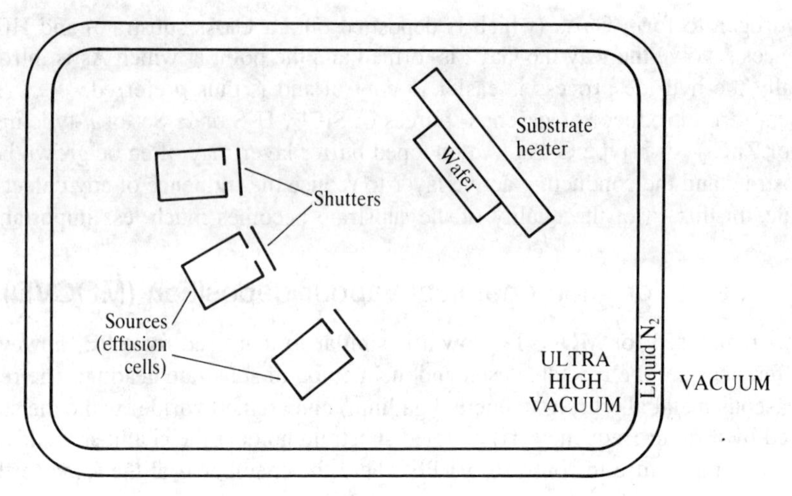

Figure 5.4 Schematic diagram of molecular beam epitaxy (MBE) equipment

profile is very dependent upon the mass of the implanted ions. Acceleration potentials in the range 50–150 keV are typical, and the resultant doping profile is also dependent on the implant dose and this acceleration potential. In the case of GaAs, the dopants are usually Be or Zn for *p*-type material, and Si or Ge for *n*-type material. The dopant distribution approximately follows the so-called LSS theory [60], and Figure 5.5 shows the typical profile produced by the implantation of silicon into GaAs [61].

Comparing ion implantation against epitaxial layer growth as techniques for producing the conducting region in which to fabricate devices, ion implantation has the following advantages:

1. the uniformity (across each wafer) is better (excluding MOCVD and MBE which are not high volume production processes at present);

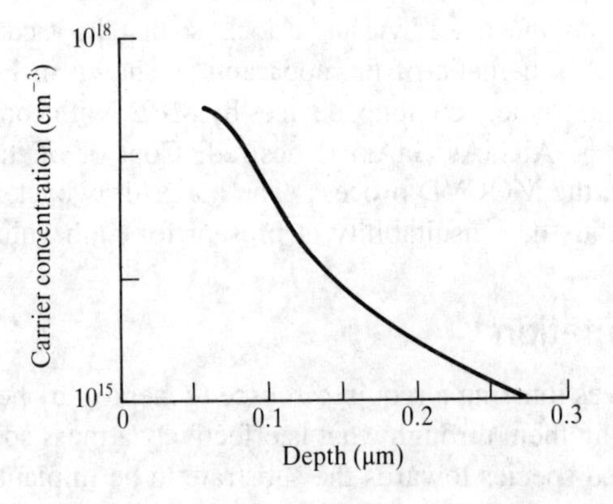

Figure 5.5 Typical carrier concentration profile for ion implantation

2. the reproducibility (from wafer to wafer) is better (excluding MOCVD and MBE which are not high volume production processes at present);
3. ion implantation has a high process rate capability, suitable for mass production of ICs;
4. different ion doses and species can be implanted globally over the whole wafer or (through the use of suitable ion masks) in localized areas as small as the device dimensions (e.g. for n^+ areas underneath ohmic contacts to reduce the source and drain parasitic resistances);
5. the resultant structures are planar thus removing the problem of step coverage by metal tracks, and making the photolithographic definition of small features easier.

Ion implantation does have some disadvantages though, and these are:

1. the implant depth is limited (of the order of 1 μm) (although this limit is increasing as new ion implanters designed specifically for GaAs appear on the market);
2. abrupt profiles are not possible;
3. there is no heterojunction capability;
4. high quality substrates (with a low defect density and a high resistivity) are required;
5. an annealing process is needed after the implant to make the dose electrically active;
6. the annealing stage is complex in order to prevent non-stoichiometry (changes in the ratio of gallium to arsenic atoms may occur as arsenic is lost at high temperatures).

The best GaAs devices made seem to use a mixture of both ion implantation and epitaxial layer growth techniques. Vapour phase epitaxy is used to grow an undoped buffer layer into which the conducting layer is implanted; the VPE process gives a very good quality substrate while ion implantation of the conducting layer gives good device uniformity and reproducibility.

5.3 GaAs devices

The next stage after preparation of a doped layer in the GaAs substrate, is the definition of different devices. The types of device which are commonly used in GaAs analog and digital circuits are:

1. resistors;
2. capacitors (metal–insulator–metal (MIM) or reversed-biased diode structures);
3. diodes (generally Schottky contacts);
4. MESFETs (depletion and enhancement-mode); and
5. interconnecting tracks.

Under some circumstances components such as p–n junction diodes, JFETs and inductors may also be required.

Figure 5.6 shows the typical process flow used for preparing standard devices. The fabrication steps are generally carried out using standard photolithography techniques as developed for the silicon semiconductor industry. There are, however, several important aspects where GaAs processing differs from silicon processing. These aspects are described in the following sections.

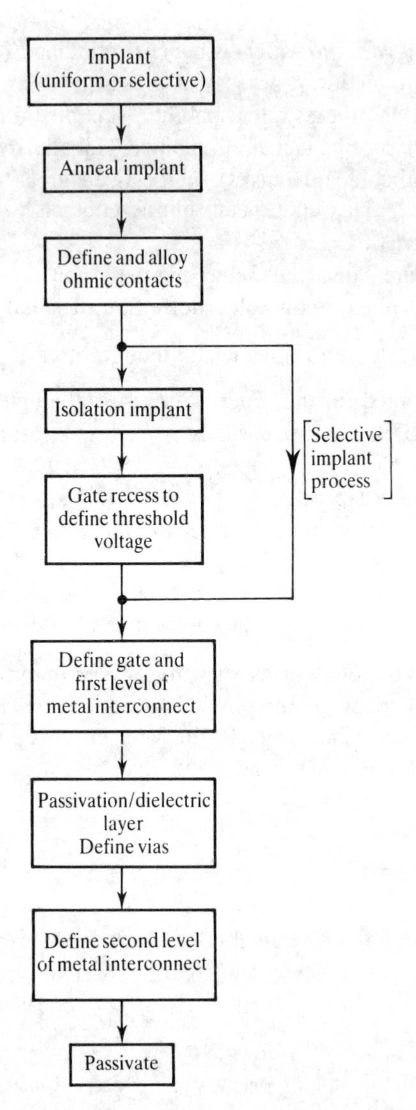

Figure 5.6 Typical process flow for GaAs IC fabrication

5.4 Photolithography

The definition by photolithography of features in GaAs ICs causes special problems [62] due to the small dimensions used (for example, submicron gate lengths are now common). Contact lithography may be used, but good contact is essential as deleterious effects caused by diffraction of the exposing radiation at these small features will be emphasized in areas of poor contact. Projection steppers may be used instead, but generally the resolvable feature size is not as small as with contact printing, and the machine cost much higher. Both methods put quite strict requirements on the wafer flatness (less than 5 μm variation over a 2 in. wafer), and planar device structures will obviously be preferred.

Optical lithography is usually carried out using mid-UV radiation with a wavelength of 300 nm, this can produce submicron feature sizes down to 0.5 μm. For subhalf micron features, deep-UV radiation (200 nm wavelength) can be used. However, an alternative technique for defining very small features (0.1 μm) is to directly write onto a resist on the wafer using an electron beam. This method has a good depth of field (about 25 μm) which reduces the flatness criterion and eliminates the significant problems associated with manufacturing subhalf micron masks, but it is not yet a high throughout production process. The use of this technique is sure to develop, but may be reserved for the definition of very small (or critically aligned) features.

When defining features made from metal in GaAs ICs, a 'lift-off' process is used rather than an 'etch' process. Both these processes are compared in Figure 5.7. The lift-off process is preferred chiefly because multiple-metal systems are commonly used for contacts, and GaAs is susceptible to attack by the many etchants which would be required to remove

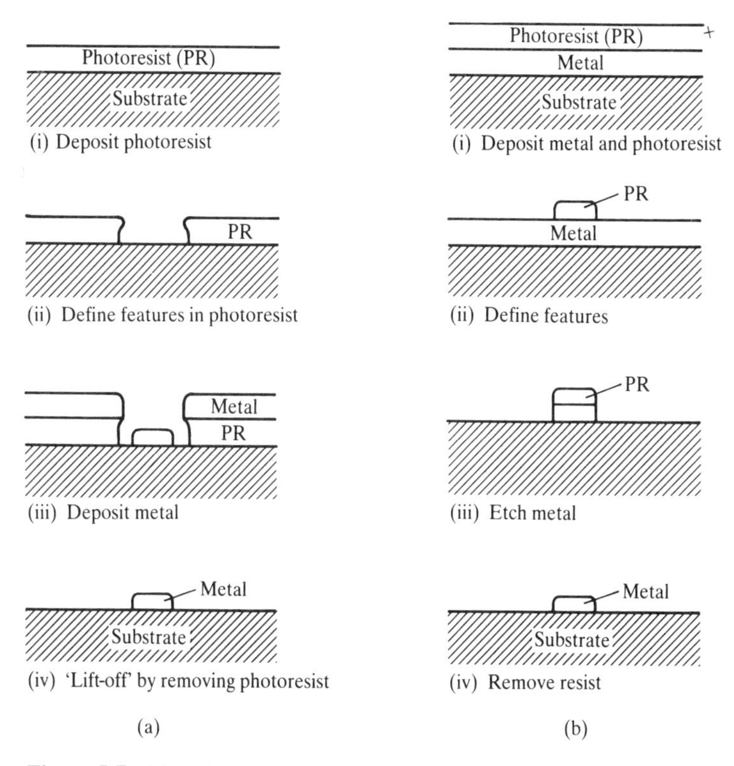

Figure 5.7 Photolithographic processes for metallization: (a) lift-off; (b) etch

these systems. The linewidth control given by the lift-off process is also better than that given by an etch process.

For the lift-off process to be successful, it is essential that the resist features have an undercut profile; this profile gives a good break between the feature and the excess metal, and prevents step coverage. The profile can be achieved by baking the resist, or by soaking in chlorobenzene [63]. Both methods effectively harden the surface of the resist by removing excess solvents. When developing is carried out the underlying resist is attacked more than the surface layer and the required undercut profile is achieved as shown in Figure 5.7(a).

5.5 Device isolation

Historically, the first GaAs ICs used a 'mesa-structure' for the isolation of devices. In this process isolation between devices is simply achieved by an etch process which leaves the doped device areas standing on 'mesas' on top of the semi-insulating substrate (see Figure 5.8(a)). Although very good isolation is obtained, this structure does cause problems

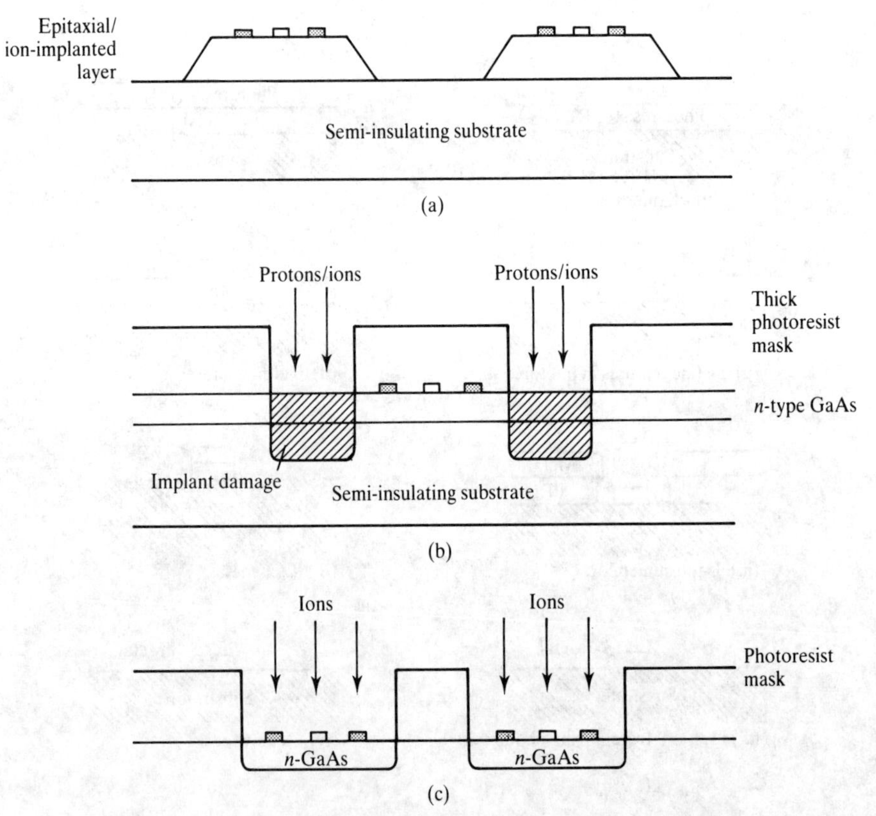

Figure 5.8 Device isolation: (a) mesa-structures; (b) isolation implant; (c) selective implantation

when it comes to ensuring good step coverage by interconnecting tracks. Generally, planar surfaces are preferred in IC production as this improves the circuit yield.

If the doped layer is produced by the implantation of a uniform ion dose over the whole wafer, then isolation between devices can be achieved by selectively implanting the wafer between the required device locations with either protons, boron or oxygen. The damage created by this implant converts the implanted n-type region back to semi-insulating material. A thick photoresist layer over the device areas acts as a protective mask in this process (see Figure 5.8(b)). Protons are easiest to mask in this isolation implant. However, the damage may then be annealed out at relatively low temperatures (typically greater than 250°C), so this isolation implant must follow all the high temperature process steps (such as alloying the ohmic contacts) in the IC fabrication schedule.

An alternative method of providing device isolation is to move from a process which involves a global n-type implant followed by a selective isolation implant to a process involving a selective n-type implant only. The material between devices is masked so that it remains semi-insulating after the conducting layer implant, and thus automatically provides device isolation (see Figure 5.8(c)). This process of selectively implanting small areas of the wafer may be extended to different doses so that, for instance, N^+ areas may be defined underneath ohmic contacts so as to reduce the parasitic contact resistance in a device, or a p-region may be placed deep under the gate to reduce short-channel effects (see Chapter 2). Selective implantation involves more difficult processing techniques than selective isolation due to the increased difficulty of masking high mass ions. However, this technique gives the significant advantage of allowing self-aligned devices to be fabricated (see Section 5.10.2) which enhance the device properties by reducing the parasitic resistances.

5.6 Annealing

After any implantation step it is necessary to electrically activate the dopant ions and to remove any damage caused by the implant. This is performed by an annealing stage. However, some means must be found of preventing the loss of arsenic from the GaAs surface during this high temperature anneal (typically carried out at 850°C for 10–30 min) as arsenic is highly volatile at elevated temperatures.

There are generally three methods used for the annealing of implants in GaAs. These methods are as follows:

1. Capped annealing: In this method the wafer is capped with a coating able to withstand the high anneal temperature. This coating prevents the loss of arsenic from the GaAs surface. Generally silicon nitride [64] (or perhaps aluminium nitride) is used.
2. Capless annealing: In this method two wafers are placed face to face in the annealing furnace. Each wafer effectively provides a proximity cap for the other, and generates an overpressure of arsenic which prevents substantial decomposition [65]. The capped annealing method always gives the possibility that the cap will leave undesirable impurities on the surface, or that stress due to different thermal expansion coefficients

will cause undesirable effects such as the introduction of defects into the crystal lattice. Capless annealing does not have these disadvantages, although the annealing environment must be more carefully controlled, particularly in respect of the oxygen and water vapour content.

3. Recently, the process of flash annealing [66] has come to be widely used. The GaAs wafer is heated to high temperatures (approximately 1000°C) in a very short time by several high intensity infra-red flash tubes. Annealing can be carried out in very short times (seconds), and this short anneal time firstly reduces any diffusion of the implanted dose, and secondly reduces the requirements on capping. Any decomposition of the wafer which may occur can be substantially eliminated by the use of an arsenic overpressure which does not need to be well controlled.

5.7 Contacts and interconnect

Ohmic contacts to GaAs are typically made using the Au–Ge–Ni metal combination, which is alloyed on the wafer at approximately 450°C. The AuGe layer is typically 1000Å thick, topped by a few hundred angstroms of nickel. These contacts allow contact resistances of approximately 10^{-6} Ω cm to be consistently achieved.

Schottky contacts are not alloyed, and are typically made using either Ti–Pt–Au or Ti–W–Au. The titanium in these systems provides a good high barrier Schottky contact, but would give a high parasitic gate resistance if used alone. This parasitic resistance is reduced by the top layer of gold. However, gold is an amphoteric dopant which would convert the Schottky contact to an ohmic contact if it came into contact with the GaAs. The middle platinum or tungsten layer is thus provided to prevent this conversion by acting as a barrier which prevents the diffusion of gold towards the GaAs surface.

Interest has also been focused on various silicide compounds, in particular tungsten and TiW silicide [67]. These silicides give useful low gate resistance, and also retain their Schottky characteristics at high temperatures which is an essential requirement for some self-alignment processes (see Section 5.10.2).

To interconnect the devices in a GaAs IC, generally at least two levels of metallization are required to allow cross-overs. In these cross-overs it is important that the two tracks are electrically isolated from each other; this can easily be achieved through the use of an intervening dielectric layer (see Section 5.8). Where connections between the two metallization layers are required then 'via' holes can be etched through the dielectric at appropriate locations (see Figure 5.9). The upper ('second') level of metallization is often TiAu, although PtAu or NiAu have also been used. This level is often used for power buses, so greater thicknesses are used (typically 0.7–1.0 μm) than on the first level (typically 0.3–0.5 μm thick). The first level may be formed using the same metal system used to create the Schottky gates, and laid down simultaneously. However, there is some advantage to be gained if the first level of metal interconnect is defined in a separate process step since this allows the use of a metal system (typically TiPtAu) aimed at minimizing the series resistance, while the Schottky contact process step is optimized independently using a different metal system (e.g. TiW silicide).

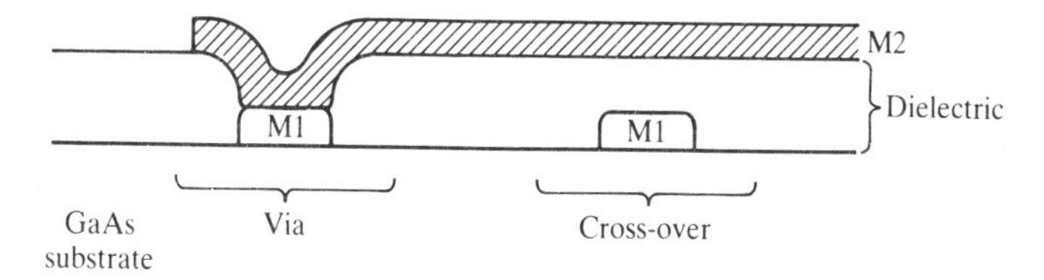

Figure 5.9 Schematic diagram showing cross-overs and via connections between two levels of metallization

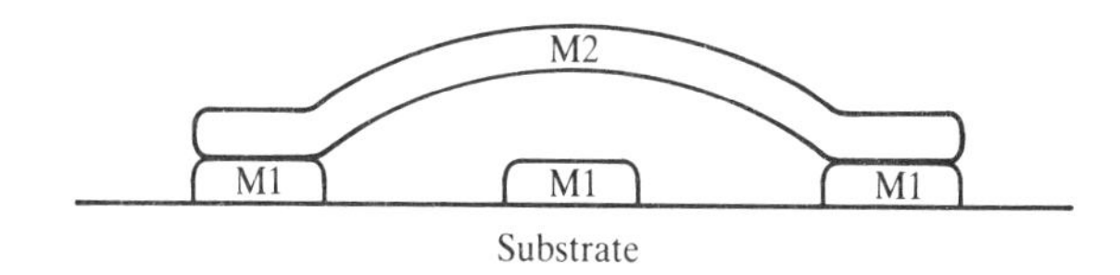

Figure 5.10 Schematic diagram of an air bridge

In some applications it may be desirable to minimize the parasitic capacitance associated with track cross-overs. This can be achieved using a structure known as an 'air bridge'. This structure uses a self-supporting second level of metallization to cross-over the first without any intervening dielectric layer (see Figure 5.10). Air bridges are fabricated by first defining the lower track in metal, then defining a lower resist pattern over this to define the bridge span. The upper track is then deposited and defined using a second resist layer. Removing both resist layers results in the self-supporting 'hump-back' bridge structure shown.

5.8 Passivation and dielectric layers

Passivation coatings are chiefly required in silicon technology to stabilize the properties of the wafer surface since, for example, surface impurities or variations in the ambient over the IC may cause inversion of the semiconductor surface and allow appreciable surface leakage currents to flow. Further undesirable effects may occur if resistances (parasitic and otherwise) are allowed to vary in the circuit. Effective passivation is given in silicon IC processing by a silicon dioxide layer.

With GaAs, a high density of energy states exists at the surface which generally makes this type of passivation unnecessary. However, some coating is required to protect the surface from possible etchants or oxidizing agents (e.g. water vapour) in the environment. This coating must be insulating to maintain device isolation, and may thus also be used as the dielectric in metal–insulator–metal (MIM) capacitor structures and the insulator between first and second levels of interconnecting tracks. Although it is common to place the first level of metal interconnect directly onto the semi-insulating substrate, a dielectric layer may also be used under this interconnect to give some reduction in the sensitivity to transient radiation [2].

Two types of passivation/dielectric coating are commonly used: silicon nitride and polyimide. Since the deposition of this coating is one of the last in the IC production schedule, high temperatures which could spoil the properties of ohmic or Schottky contacts must be avoided. Silicon nitride is preferred when high dielectric constant films are required (capacitor areas will be smaller). Deposition is by plasma-enhanced chemical vapour deposition (PECVD), whereby ammonia and silane (SH_4) are reacted in an rf-induced plasma at 300°C. Polyimide is a polymer coating which is preferred for its ease of application (it can be spun-on in a manner similar to the application of photoresist); its planarizing property, which reduces step-coverage problems if a second-level metal interconnect is used; and for its low temperature curing requirement (200°C–400°C). Care should be taken that this curing is correctly achieved, otherwise retained water may give capacitors a high dielectric loss and cause some circuit problems, particularly at high frequencies. Via holes to allow connections to be made between first and second levels of metal interconnect are usually made by etching with a reactive gas or plasma at low pressure.

5.9 Threshold voltage determination

The threshold voltage of a MESFET is the gate bias at which the Schottky depletion layer just extends through the conducting channel to the semi-insulating substrate (see Chapter 2); at gate biases more negative than this the channel is cut off and channel currents cannot flow. The threshold voltage is thus a function of the diode built-in voltage (which determines the initial depletion layer position), the channel doping density and the channel thickness. Other factors, such as the number of defects in the wafer and the parasitic resistances, will also exert a (smaller) influence on V_T.

For LSI circuits it is essential that the threshold voltage is well defined and uniform if reasonable yields are to be attained (see Chapter 3). To manufacture MESFETs with a predetermined threshold voltage two major methods are used. These are based upon the following features:

1. The use of a pre-calibrated implant. The doping level and channel thickness are dependent on the implantation conditions, hence the threshold voltage can be controlled by varying these conditions.
2. Etching the conducting layer which will be present under the gate until a known thickness results and hence a known threshold voltage. This 'gate recessing' can be carried out, for example, by monitoring the current flowing between two contacts, and stopping the etch at a predetermined position. An alternative solution is the use of the 'self-limiting anodic etch' process [62] (see Figure 5.11).

In the anodic etch process, electrical contact is made to the conducting layer in the wafer which is then placed in an electrolytic solution. If the wafer is given a positive bias with respect to an electrode placed in the solution, it becomes a reverse-biased Schottky diode with a depletion layer thickness dependent on the applied bias. Under illumination, avalanche breakdown can be induced in this depletion layer at high voltages (typically

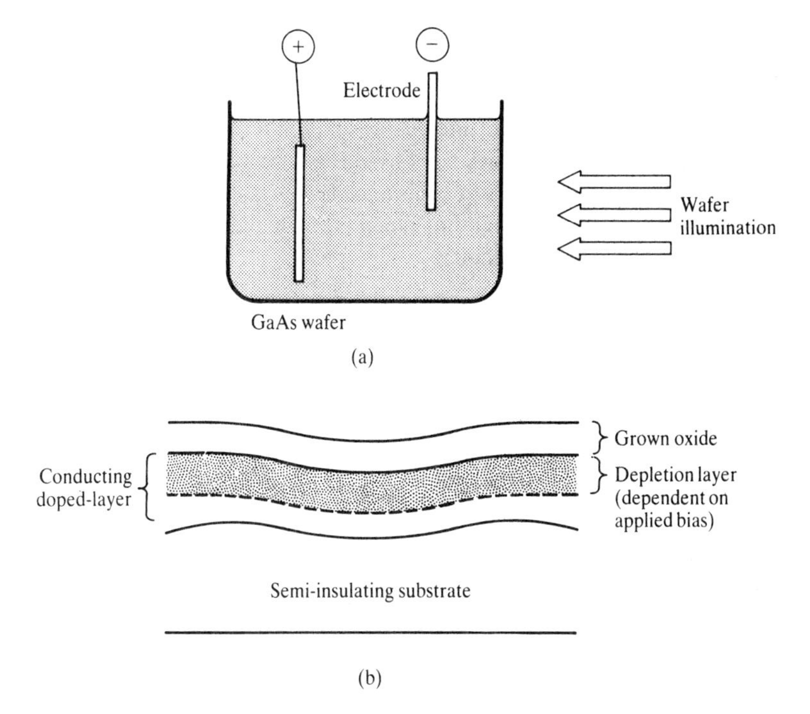

Figure 5.11 The self-limiting anodic etch process: (a) anodization cell; (b) cross-section through GaAs during anodization

voltages of the order of 50 V are used), and the resulting current will allow electrolytic oxidation (anodization) to take place at the wafer surface. As the oxide layer at the surface grows, then so the position of the depletion layer is 'pushed' deeper into the substrate. Eventually the semi-insulating substrate will be reached, at which point electrical contact to the channel will be lost and anodization will stop. This condition can be monitored by recording the anodization current and removing the wafer when the current suddenly drops (by many orders of magnitude).

At the end of anodization the depletion layer due to the applied bias will meet the substrate interface over all the wafer, independent of any local variations in doping level or channel thickness. Thicker or more highly doped regions will simply anodize for longer before the depletion layer reaches the semi-insulating substrate. Since the threshold voltage is determined by the condition that the depletion layer extends to the substrate, the threshold voltage of devices on any part of the wafer will be essentially equal, overcoming any initial wafer non-uniformities. (Any threshold voltage variations will be essentially determined by smaller, second-order effects such as parasitic resistance variations due to lithography tolerances.) Control of the final threshold voltage is simply achieved by varying the wafer bias used in the anodization cell. The oxide layer is easily removed by etching in dilute hydrochloric acid.

5.10 Reduction of parasitic source and drain resistance

In the basic MESFET structure, parasitic source and drain resistance arises from contact resistance and from the channel resistance of the region between the gate and the ohmic contact. It is desirable for this parasitic resistance to be small to reduce any degradation of the device transconductance and circuit speed (see Chapter 2). It is generally reduced by using one of two structural methods:

(a) recessed gate structures, or
(b) self-aligned structures.

These methods are described below.

5.10.1 Recessed-gate structures

By using a thick channel layer to start with, and selectively thinning this layer only under the gate electrode, devices with any required threshold voltage can be fabricated while maintaining low parasitic resistances. This selective thinning (recessing) of the area in which the gate is placed (see Figure 5.12) is generally achieved by one of three methods:

(a) a wet chemical etch process, e.g. H_2O to oxidize the surface and NH_4OH or H_2SO_4 to remove this oxide;

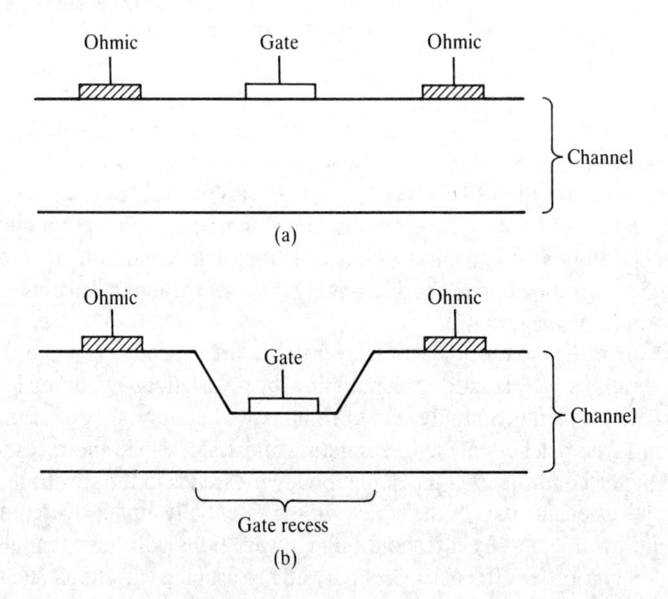

Figure 5.12 Comparison of planar and recessed-gate MESFET structures: (a) planar; (b) recessed-gate

(b) a dry etch process using a reactive gas or plasma; or
(c) the self-limiting anodic etch described in Section 5.9. As noted before this method gives the advantage of improving the threshold voltage uniformity in the event of wafer non-uniformities.

Since the regions either side of the recess are thicker, the parasitic resistances present in devices with a recessed-gate will be correspondingly lower than in devices with a planar structure with the same threshold voltage.

5.10.2 Self-aligned structures

Reduction of the parasitics in a self-aligned process comes chiefly from maintaining a very small separation (typically 0.25 μm) between the gate and a low resistance, N^+ contact area under the ohmic metallization. This small separation could not be reproducibly maintained by standard photolithography techniques which use a different mask for each feature, so one mask has to be used to define the position of both the gate and the N^+ features. The process of effectively using one feature to automatically align another is known as self-alignment. In GaAs MESFET production, two self-alignment processes are predominantly used; a gate-priority scheme and an N^+ priority process.

In the gate-priority process, the gate is given a T cross-section, and the top of the T acts as a mask to maintain separation between an N^+ implant and the bottom of the T (which defines the gate length). This separation is defined by the amount of undercut allowed when fabricating the T cross-section, and hence this step determines the magnitude of the parasitic resistances. Since the gate is present during the implant step, it will also be present at the implant anneal. This self-alignment procedure, although conceptually very simple, is thus restricted to the use of Schottky contacts which retain their characteristics at high temperatures. TiW silicide [67] and W–Al alloys are possibilities. Figure 5.13 shows the essential steps in this 'stable gate' self-alignment process.

In the N^+ priority process (otherwise known as SAINT for self-aligned implantation for N^+ layer technology) [68], the processing is more complex but not dependent on finding materials which retain their properties at high temperatures. Separation between the N^+ areas and the gate is again defined by a T-shaped structure, but this time the T is made of photoresist and silicon nitride. The top of the T again acts as a mask for the N^+ implant, while the bottom defines the gate position, but the gate is not deposited until after the implant anneal. This allows conventional metallizations to be used. The essential steps in the SAINT process are shown in Figure 5.14.

In order to act as an efficient mask to the N^+ implant, the photoresist T must be very thick. If standard photolithography were used to define this structure then it would be very difficult to maintain the required profile due to the isotropic action of developers; with thick resists the normal tendency is to form structures as shown in Figure 5.15(a). In order to achieve the required profile a tri-layer of thick photoresist, silicon nitride and then thin photoresist on top is used as shown in Figure 5.15(b). The thin photoresist is used to transfer the N^+ implant mask dimensions to the silicon nitride, and then a highly anisotropic etch process is used to etch the thick photoresist. Typically reactive ion etching

1. Selectively implant n-layer
2. Deposit stable gate metallization
3. Deposit and define metal mask

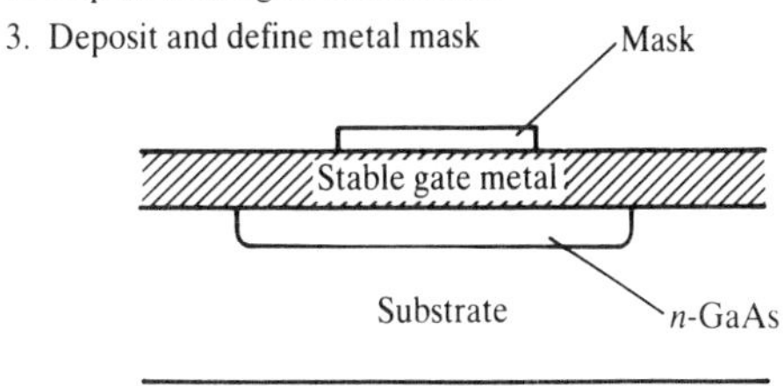

4. Etch (and undercut) gate metallization

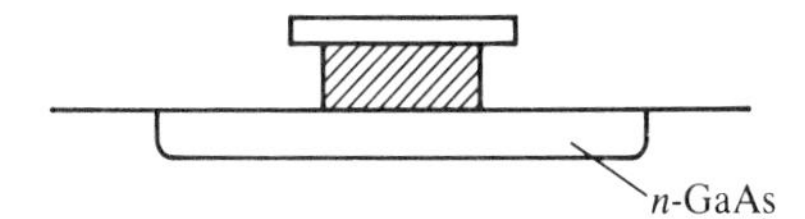

5. Implant N^+ areas

6. Remove metal mask
7. Cap with silicon nitride and anneal
8. Etch windows in nitride cap, then deposit and define ohmics (and alloy)

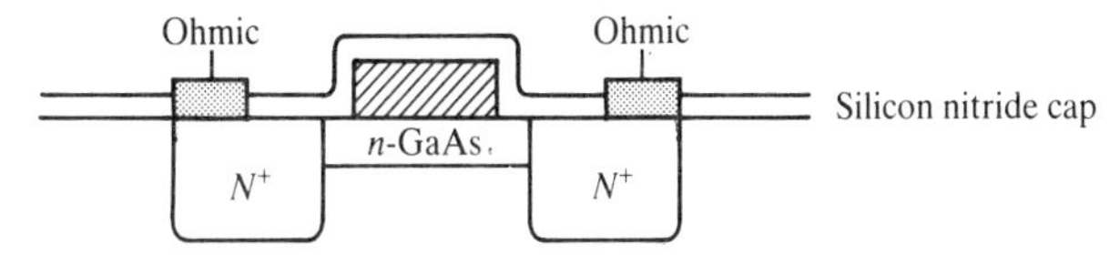

Figure 5.13 The 'stable gate' self-alignment process

(RIE) is used whereby the etch rate of reactive ions is given a high degree of directionality by accelerating the ions perpendicular to the substrate with a negative bias. The silicon nitride layer protects the top of the thick photoresist (and thus maintains the mask thickness) during this anisotropic etch.

5.11 Packaging

The packaging of a GaAs IC must fulfil two main requirements: firstly, the package must allow dissipation of the heat generated in the IC; and secondly, the package must allow the IC to operate with its full speed performance.

Dissipation of the heat generated in the IC is dependent not only on the package characteristics, but on the transfer of energy from the devices in the IC through the substrate to the package. The thermal conductivity of GaAs is only 60 percent of that of silicon,

1. Selectively implant n-layer
2. Coat wafer with silicon nitride, then tri-layer of thick photoresist/silicon nitride/thin photoresist
3. Define N^+ areas in thin resist layer then etch in nitride

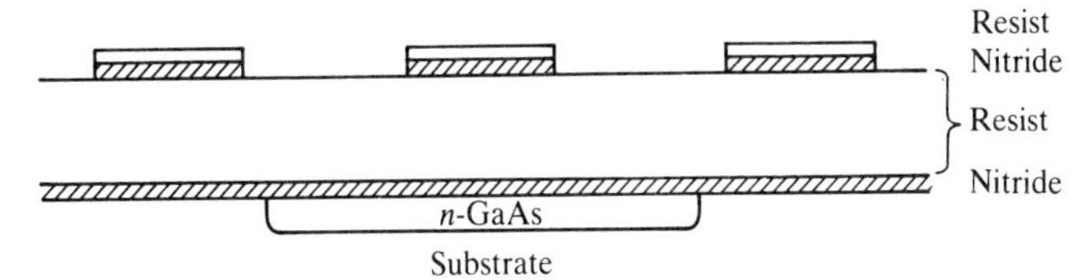

4. Use anisotropic etch to define N^+ mask in thick resist
5. Implant N^+ areas

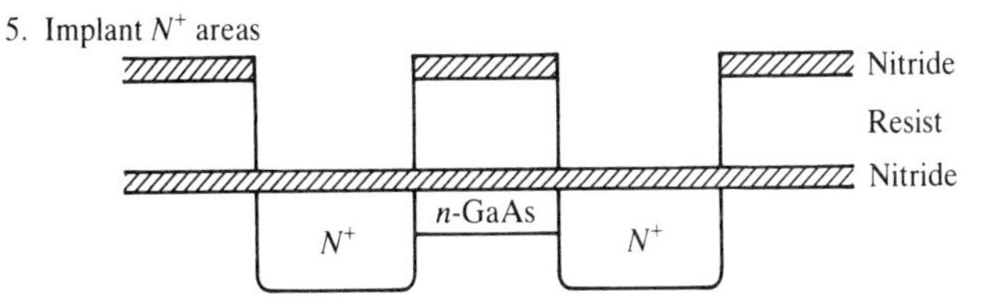

6. Undercut resist (define T-structure) then deposit dielectric layer (silicon dioxide)

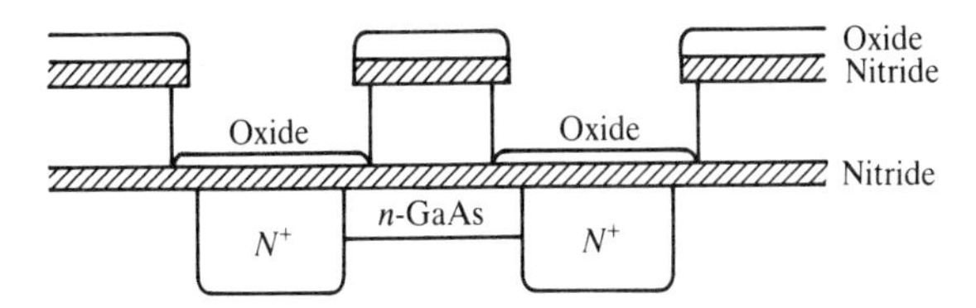

7. Remove photoresist and anneal implants

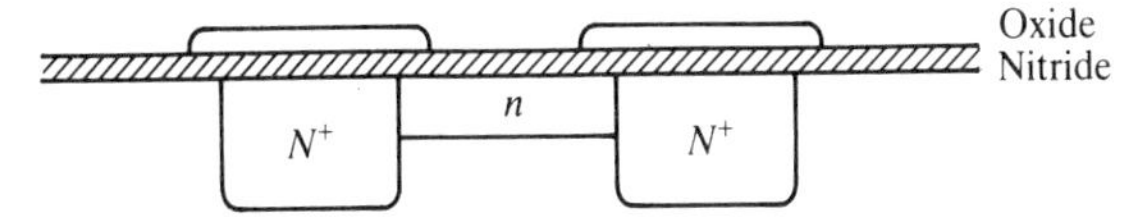

8. Etch windows in oxide and nitride for ohmic metallization
9. Deposit ohmics, define and alloy
10. The gate window is etched in the nitride using the oxide as a mask, then gate metal is deposited and defined

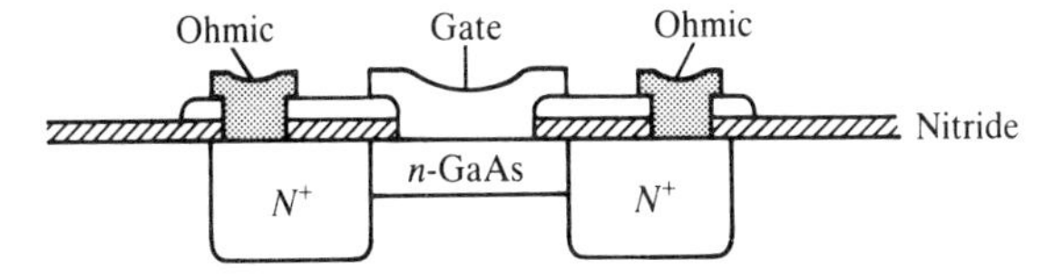

Figure 5.14 The SAINT process

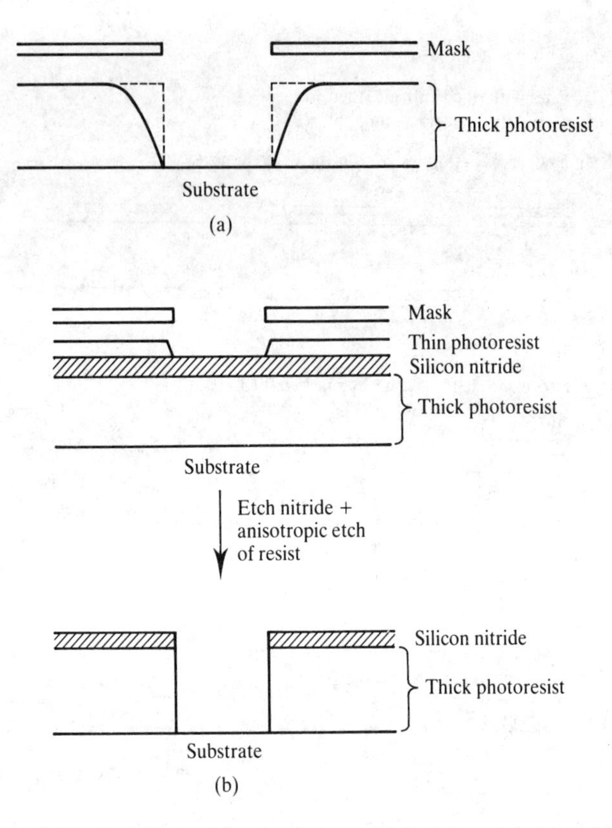

Figure 5.15 Definition of features using a thick photoresist: (a) resist only; (b) tri-layer structure

and thus it is relatively more difficult to achieve this energy transfer. A GaAs IC will thus run hotter than a silicon IC with the same power dissipation (although the wider bandgap of GaAs makes it less sensitive to high temperatures). At high power dissipations it may be desirable to improve the energy transfer by reducing the thickness of the wafer using a chemical etch or a mechanical lapping process. The processed wafer has a thick layer of wax placed over the devices to provide protection during this thinning process; the wax can be removed easily prior to packaging. The original thickness of a GaAs wafer is typically 350 μm, with care this can be reduced to perhaps half this value.

Obtaining full speed performance from the packaged IC is dependent upon several factors. Firstly, the series inductance on all lines into and out of the IC should be minimized. A package with a well size which allows minimum bond wire lengths should thus be chosen. For critical pins (i.e. the ground connection), bond wires may be connected in parallel to reduce the overall inductance. Secondly, the parallel capacitance on all signal lines should be minimized. Packages with narrow (and short) traces between the well and the outside contacts will be preferred. On power supply lines capacitance is actually helpful in terms of providing a good, clean supply voltage; for ultimate speed performance it may be necessary to use custom packages having decoupling capacitors and signal terminating resistors built into the package (see Chapter 6). The last requirement concerns grounding of the substrate. Although intrinsic GaAs is insulating, very small (picoampere) currents can circulate in the substrate and affect the parameters of the GaAs MESFETs; in particular low frequency oscillations may be induced [69]. Connecting the substrate to ground by a large area contact on the back of the IC helps to suppress these oscillations by providing a path to ground for any substrate currents.

Finally, note that wire bond connections to silicon ICs are generally made by one of two techniques: thermocompression bonding or ultrasonic bonding. The latter technique is not recommended as a means of bonding to GaAs due to its high fragility.

6 □ Layout and system considerations

The earlier chapters in this book have described in turn the characteristics of GaAs devices, circuit design techniques for digital and analog ICs, and the technology required for fabrication of these devices. At various points these chapters mentioned aspects which must be remembered when generating the integrated circuit layout data which will be used in the fabrication process. This chapter has two aims: firstly, to bring these various aspects together with a set of layout rules for a generic GaAs IC processing technology, and secondly, to describe techniques which allow the potential high intrinsic performance of GaAs to be exploited properly within a complex electronic system.

6.1 Layout rules

The layout of an IC describes the shapes present in the set of masks which are used in the IC fabrication process. Each mask is associated with a different step in the fabrication process of the IC, and the shapes define the positions of the features produced by the fabrication step. A process step cannot produce features of an arbitrary size in arbitrary positions while still yielding devices with desirable characteristics and a good circuit yield. Generally there are constraints on how small and how closely spaced features may be, and each process therefore has an associated set of layout rules which define the minimum sizes of the features on each mask and also the minimum clearances to other features on the same or other masks. This section illustrates these layout rules as applicable to a generic non-self-aligned process.

In an attempt to give a degree of manufacturer-independence, these layout rules have been described in relation to the parameter λ, which in this case is defined to be the minimum resolvable feature size. It is also assumed that features on different masks can be aligned within a tolerance of $\pm\lambda$. In the examples given here, λ is assumed to be one micron, and most manufacturers of GaAs ICs will be able to fabricate ICs based upon these rules. It is important to emphasize, however, that in order to be generic, these rules are also conservative (for example, many manufacturers now routinely process devices with a gate length of only 0.5 μm). Higher performance will be obtained if the rules given for minimum size and separation are modified according to the specific requirements of the chosen IC manufacturer.

The generic layout rules are illustrated in Figure 6.1. It is assumed to use selective implantation to define both enhancement and depletion-mode MESFETs with the required threshold voltages. Separate masks are thus required to define the channel implant regions in devices with different threshold voltages, and the channel implant feature of a MESFET

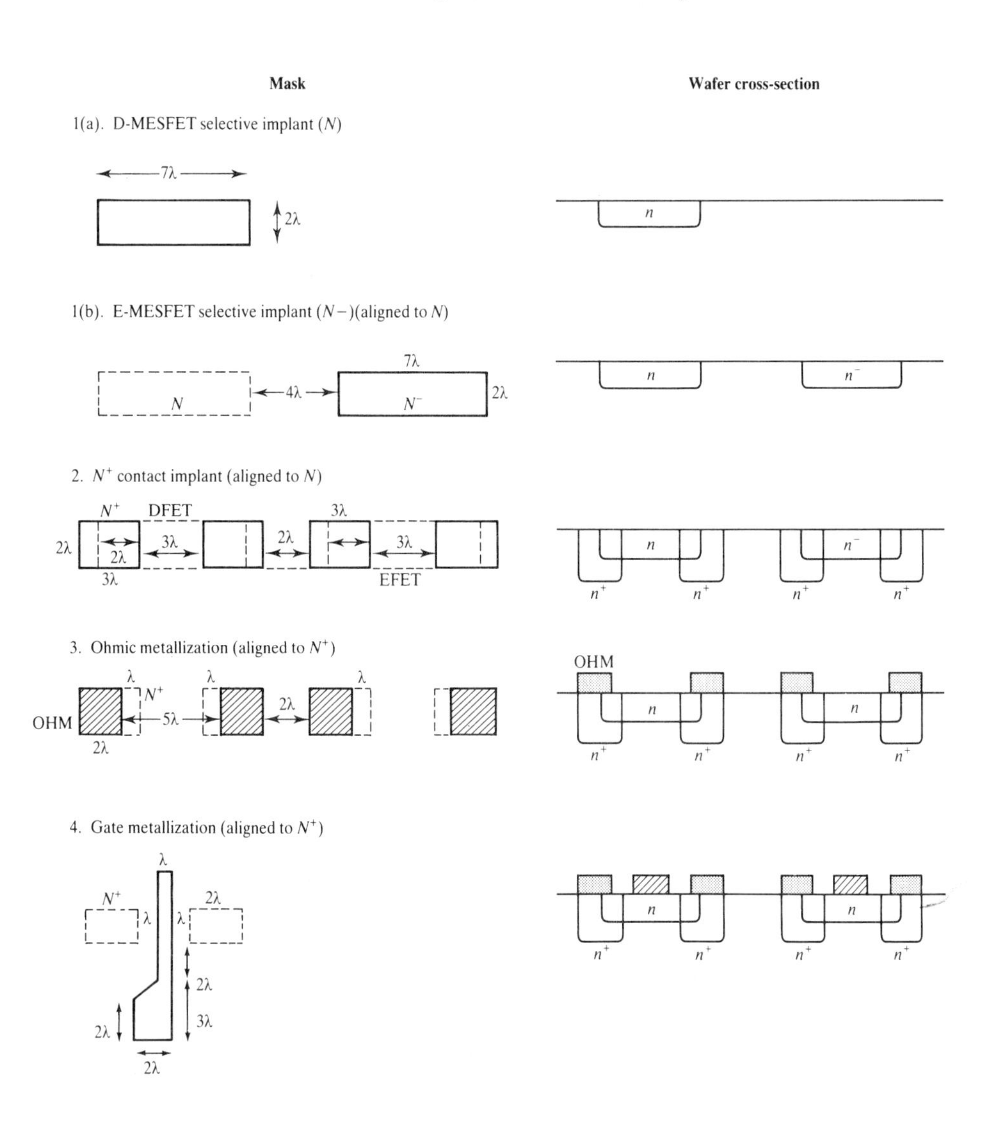

5. Metal 1 (aligned to GATE)

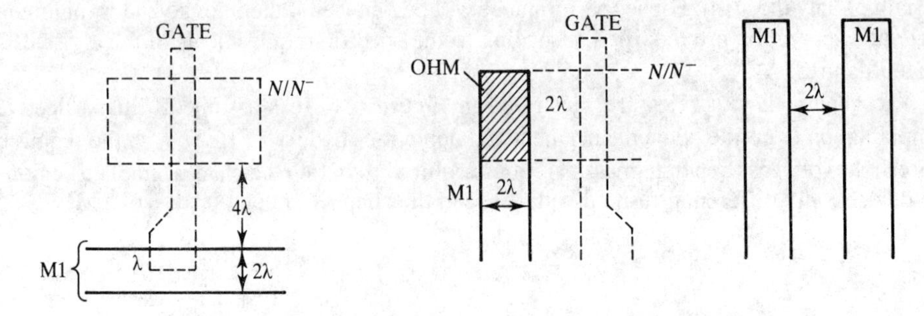

6. Dielectric via holes (aligned to M1)

7. Metal 2 (aligned to VIA)

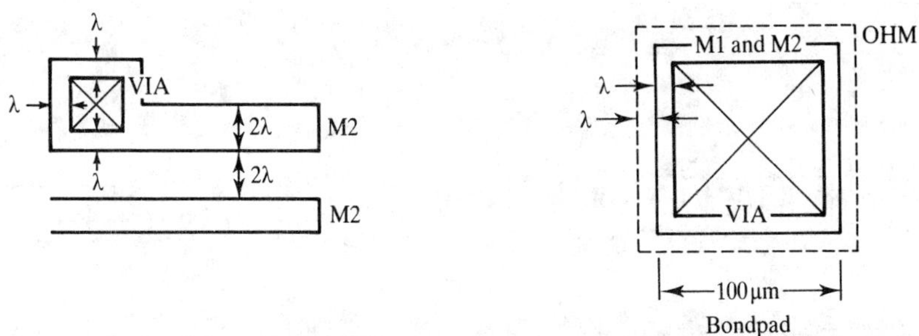

Figure 6.1 Generic layout rules for a non-self-aligned selective implant GaAs MESFET process

(or a diode) should appear on the mask appropriate to the required implant. An n^+ selective implant is also used underneath the ohmic contacts to reduce the ohmic contact resistance. Two levels of metal interconnect are available, separated by a silicon nitride dielectric layer (which can also be used as the insulator in MIM capacitor structures). The process is assumed to use different metallizations to define the Schottky gates and the first level of metallization. The processing may be simplified by using the Schottky metal for both purposes, although it should be noted that this may lead to high values of parasitic resistance in the interconnect with some metal systems (e.g. silicide-based contacts in particular).

6.2 Layout considerations

The number of masks and the exact purpose of each within an IC fabrication schedule will depend on the process involved (e.g. whether it is non-self-aligned or self-aligned), and also on the particular requirements of the manufacturer. The layout rules generally specify the minimum parameters which must be met when designing the layout. However, the fact that these rules have been obeyed does not automatically mean that a working IC will be fabricated, particularly if the IC is intended to operate at high frequencies. There are a number of further considerations which should be taken into account when designing the layout of a GaAs IC. These considerations are described in the following sections.

6.2.1 Parasitics in the supply rails

At the circuit design stage the interconnections between devices are assumed to be ideal with zero impedance. In practice the interconnections on an IC are not ideal but possess finite resistance, capacitance and inductance, which are dependent upon the metal resistivity, thickness, width and topology. Where the interconnections are used as supply rails, the parasitic capacitance can be desirable since it helps to 'smooth' any variations in the supply voltage. Within an IC the parasitic inductance can usually be neglected. However, the parasitic resistance can produce a non-negligible potential drop when current flows so that different parts of the circuit see different supply voltages. In analog circuits where the power rail is used as a reference voltage, it may be necessary to provide a separate reference track for those situations where appreciable current flows in the supply track. In digital circuits such as shown in Figure 6.2, the supply track parasitics will cause logic gate Z to see a reduced supply differential compared to gate A. This will result in a reduced output swing, raising the logic low level and lowering the logic high level, and thereby eating into the noise margin of the circuit. Digital layouts must thus be defined so that the potential drop developed across the supply rail resistance is insignificant compared with the available noise margin. This is particularly important along the ground or V_{SS} rail (especially in DCFL circuits) since the logic low noise margin is often the parameter which limits the circuit yield. As a rough guide, no more than 10 mV should be allowed to develop along the ground rail. This aim can be met firstly by scaling the width of the supply track in line with the sheet resistivity of the track metallization and the anticipated level of current,

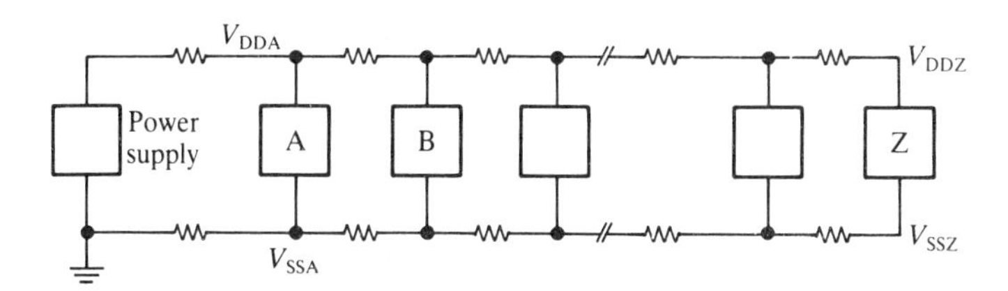

Figure 6.2 The effect of parasitic resistance in supply rails

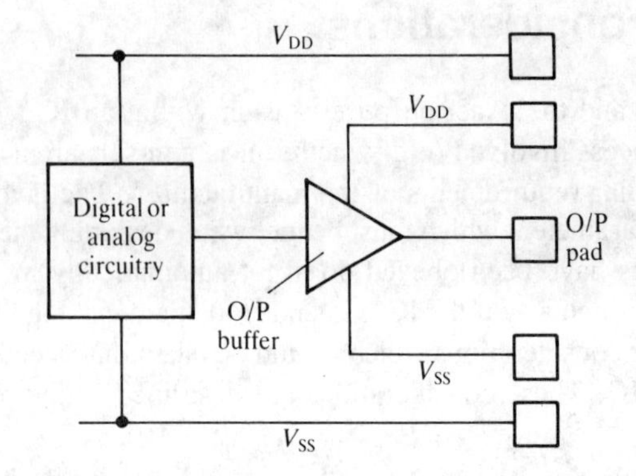

Figure 6.3 Power supply distribution to output buffers

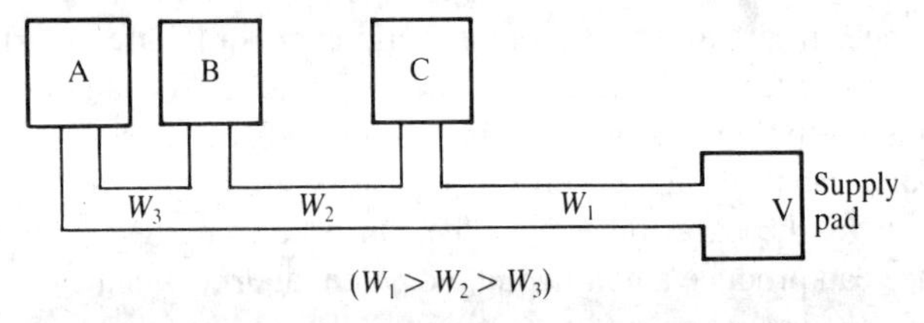

Figure 6.4 Power supply distribution to internal circuitry

and secondly by providing separate (short) supply rails for those parts of the circuit involved in switching large currents (e.g. the output buffers) (Figure 6.3). The area required can be minimized if a 'tree-like' structure (Figure 6.4) is adopted so that the track width decreases (in line with the current in the track) with increasing distance from the supply bondpad, provided that the total potential difference developed along the track remains small.

6.2.2 Parasitics in signal lines

On the signal lines the levels of current and length of interconnect are generally much less than on supply tracks. Parasitic inductance and resistance thus produce much smaller potential drops and can generally be neglected. Where the signal lines are very long, it may be necessary to treat them as transmission lines as discussed later in this chapter. Ignoring this exception, however, the principal parasitic component on signal lines is parasitic capacitance. Unless this capacitance is negligible compared to the capacitance load imposed by the devices on the line, it will limit the rate of change of the signals and thus limit the maximum frequency of operation of the IC.

The track capacitance can be minimized by reducing the length and width of the interconnect, and also by increasing the separation to neighbouring tracks. This latter factor produces the required reduction in capacitance since GaAs ICs use a semi-insulating substrate so that track–track capacitance is often more significant than track–ground capacitance (which is the principal component of capacitance in silicon ICs having a

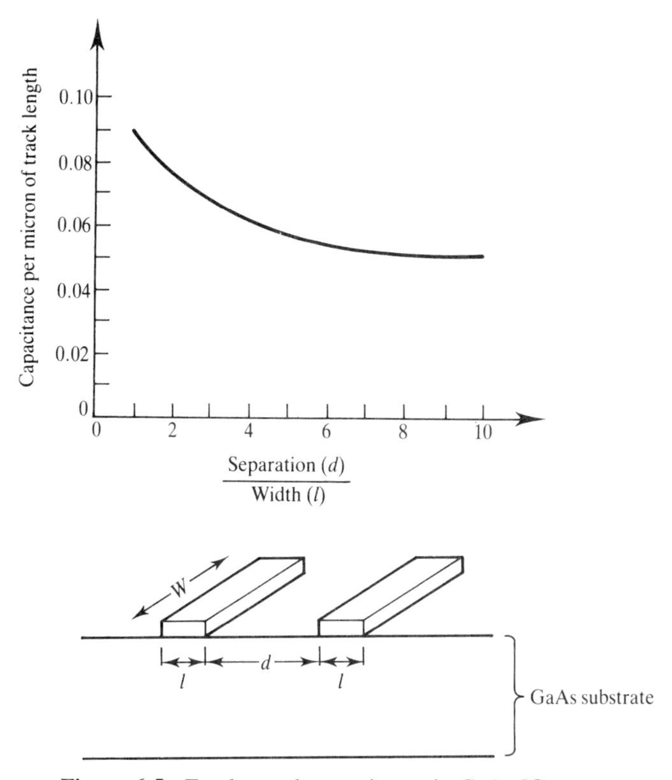

Figure 6.5 Track–track capacitance in GaAs ICs

semiconducting substrate). For a track running across a GaAs wafer of the typical thickness of 350 μm, the capacitance to ground (the bottom of the wafer) is of the order of 0.0003 fF μm^{-2}. In comparison the track–track capacitance is much greater, and of the order of 0.100–0.050 fF per micron of track length, as shown in Figure 6.5. The track–track capacitance depends upon the ratio of the track separation distance (d) to width (l), according to the empirical relationship [70]:

$$C_{\text{track}} = \frac{1.39 \times 10^{-2}(\epsilon_r + 1)W}{\ln[4(1 + d/l)]}$$

It can be seen that ratios of track separation to width of less than 3 lead to a rapid increase in the track–track capacitance. This leads to a design rule which requires long parallel lines to be separated by at least two to four times the track width.

6.2.3 Bondpad usage

In many silicon ICs the bondpads are usually positioned around the perimeter of an IC, with little restriction on their use. However, at the very high operating speeds associated with GaAs ICs (and some silicon ICs) some advantage will be gained if some restrictions are imposed. These restrictions are summarized as follows:

1. Power supply contacts should be placed in the middle of a side, rather than near to a corner of the chip. This reduces the bondwire length necessary to link the chip and the package, and thus minimizes the stray series inductance (which is typically 2.6 nH per bondwire [71]). This is desirable since reliable high speed operation depends on the availability of a constant voltage supply, and this in turn requires that the supply has a low impedance. Any series inductance will raise the impedance of the supply, so that sudden changes in the current drawn by the circuit may produce large changes in the supply potential at the chip, and these supply changes represent a source of electrical noise in the system. Since the magnitude of the voltage change depends on the rate at which the supply current changes according to:

$$v = L\frac{\partial i}{\partial t}$$

this source of noise is obviously more important in high-speed circuits such as GaAs than in standard silicon products.

   The problem of noise on the supply lines is particularly troublesome for the ground rail. In digital circuits the logic low noise margin is often much smaller than the logic high noise margin, hence the system is more sensitive to noise on the ground rail. In analog circuits the ground is often used as a reference level; by definition reference levels should be stable, not subject to large quantities of electrical noise! If possible, therefore, several parallel ground connections should be made to the chip to reduce the total bondwire inductance.
2. In sampled analog ICs, or other circuits where both high speed analog and digital signals are present, separate ground tracks and bondpads should be provided for the analog and digital circuitry. If this is not done then the current flowing to ground in the digital circuitry may produce a ground rail potential which varies in time in synchronism with the current drawn by the digital circuitry; some of this variation will be due to the potential developed across the bondwire inductance, and a further element may be contributed from the potential dropped across the finite resistance of the tracks on the IC as described above. Since the analog ground rail is often used as a reference level, currents flowing in the digital circuitry should be prevented from producing changes in the level of the analog ground. Separate analog and digital ground rails should thus be provided, and these should not be combined except at a point with very low impedance outside the chip package.
3. Interaction (or cross-talk) between different signals may be reduced if the signal inputs are restricted to one side of the IC and the outputs to the opposite side. The task of designing the p.c.b. for the system may also then be made easier. The degree of cross-talk will be worst if the input signals originate from small-signal high impedance sources, and the outputs are provided by large-signal low impedance sources: this is a common scenario!

6.2.4 Gate electrode orientation

In most GaAs ICs the mask used to define the gate electrode is the most critical since it usually has the features with the smallest dimension (the gate length is required to be small to obtain good power-delay products in digital ICs and high bandwidths in analog ICs), and also the smallest separation to other features on other masks (the parasitic source resistance is required to be small to reduce the potential degradation in the transconductance, and hence a small gate–source separation is required). In the case of non-self-aligned MESFET structures the gate and source are defined by different masks, thus a small gate–source separation can only be achieved by maintaining a close alignment between the two masks concerned. This task is considerably easier if all the devices have the same orientation (i.e. all the gates are aligned along the same axis) so that small alignment tolerances are only required along the axis perpendicular to the gate width.

Self-aligned structures have been developed so that small separations between features can be achieved without needing to maintain the accurate relative mask alignments required by non-self-aligned processes. However, there is still some advantage to be gained by aligning all gates along the same axis in terms of the reproducibility of the MESFET parameters. The threshold voltage in particular is dependent to a small degree upon the gate orientation relative to the crystal lattice [72] from piezo-electric effects [73], and restricting the orientation to one direction will give some reduction in the standard deviation of the threshold voltage; Chapter 3 showed that this is especially important if VLSI circuits in GaAs are required.

6.2.5 'Distributed' MESFETs

In some applications (e.g. analog to digital converters) it is very important that extremely good uniformity in the MESFET parameters is obtained. The intrinsic uniformity of the wafer can be improved by arranging for the MESFETs to be formed from the parallel connection of several smaller devices distributed over a relatively large area of the chip. The resulting 'distributed' MESFET will have characteristics roughly equivalent to the average of the characteristics of the component MESFETs, and thus the uniformity will be improved [74]. It is important to note that the input capacitance of the distributed MESFET will also be increased due to the capacitance of the connecting tracks; a trade-off exists between the degree of uniformity obtainable and the input capacitance. For the highest speed applications, improvements in the uniformity must come from processing improvements not layout tricks.

6.2.6 Backgating effects

The modulation of the channel current flowing in a GaAs MESFET is usually achieved by varying the bias on the gate electrode. It is also possible, however, to modulate this current by effecting changes to the number of carriers present at the epi-layer–substrate or implanted-layer–substrate interface. These changes can be made by applying relatively large (typically greater than 5 V) negative potentials to an ohmic contact positioned near to the MESFET (Figure 6.6) [75, 76]. The bottom of the channel acts as if controlled by a second gate electrode, and thus the modulation of current by this depletion layer is said to be due to 'backgating' or 'sidegating' effects. The magnitude of the backgating effect is dependent on the magnitude of the side-contact bias, the separation of the contact

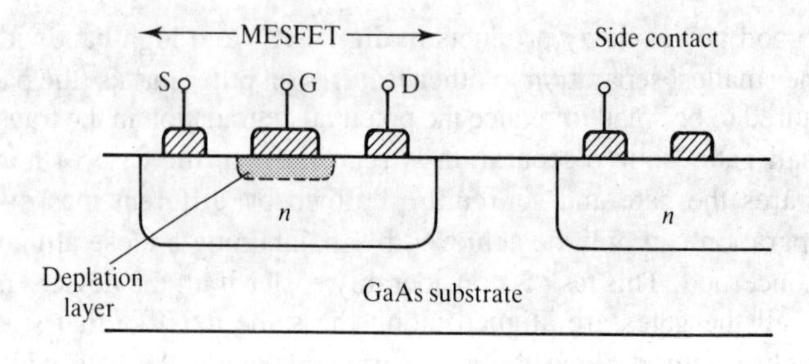

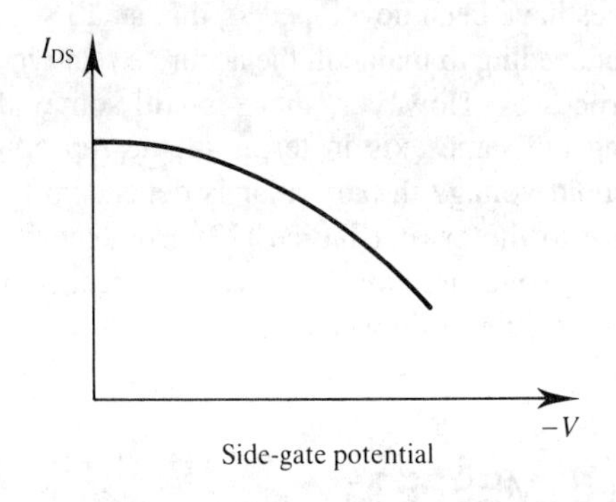

Figure 6.6 Backgating effects in MESFETs

from the MESFET, the substrate material and the method used for isolating devices (e.g. proton, boron or oxygen implantation). Increasing the side-contact bias (negatively) above a certain threshold voltage decreases the number carriers available at the active layer–substrate interface and thus reduces the channel thickness and current. The reduction in the channel thickness by backgating can be considered as equivalent to a shift in the MESFET threshold voltage in the positive direction.

The occurrence of backgating becomes a problem if the circuit operation is critically dependent upon the use of matched devices, and different MESFETs in the circuit undergo different degrees of backgating. Generally digital circuits are excluded from this category since those logic families using negative supply rails (e.g. BFL, SDFL) have large noise margins. However, the operation of many analog circuits is often dependent upon ratios of device parameters. As an example, consider the analog source–follower stage shown in Figure 6.7. The negative supply rail contact (V_{SS}) may backgate Q1 and hence produce different threshold voltages in devices Q1 and Q2; this will affect the offset voltage and the gain of the stage. Alleviating the problem depends chiefly on improving the material and isolation characteristics [77, 78]. However, planning the layout so that each MESFET sits in its own well and obeys the manufacturer's layout rules regarding device separation will also reduce backgating effects.

Considering the source–follower example again, the packing density could be improved considerably by placing Q1 and Q2 together in one well, and using a common ohmic contact to act as both the source of Q1 and the drain of Q2. However, in this case Q1 would

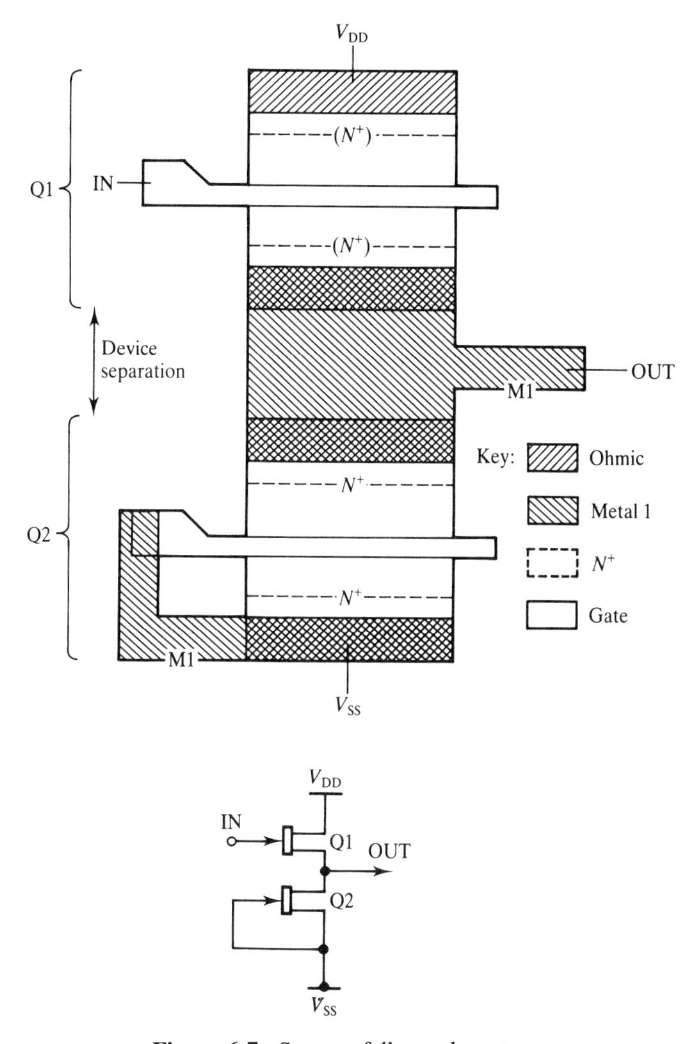

Figure 6.7 Source–follower layout

be very susceptible to backgating effects from the V_{SS} rail contact. Using well-separated wells for Q1 and Q2 can eliminate these effects.

6.2.7 Resistor layouts

When designing an IC layout it is usual to want to minimize the chip area. However, this can cause problems if it results in locally high electric fields of sufficient strength to cause saturation of the electron velocity. This is a particular problem in resistor structures where velocity saturation can cause a departure from linear I–V characteristics. This can be avoided by ensuring that the length of all resistors will keep the electric field strength below approximately 5 kV cm^{-1} (0.5 V μm^{-1}).

6.2.8 Diode layouts

Diodes are commonly used in GaAs ICs in level-shift circuits, for example, within the buffer stage in digital BFL circuits, or within analog level-shift stages. The voltage dropped across each diode is dependent upon the current through the junction according to:

$$I \approx I_s \exp\left(\frac{qV_D}{kT}\right)$$

and is also dependent upon the potential dropped across the parasitic resistance in the structure. This parasitic resistance is primarily associated with the semiconductor material between the edge of the Schottky contact and the ohmic/n^+ contact (Figure 6.8); at typical current levels this forms the major part of the a.c. resistance given by:

$$r_{\text{a.c.}} = \frac{\delta V}{\delta I} = \frac{kT}{qI} + R$$

The presence of this a.c. resistance in the diode means that as well as producing a change in the d.c. signal level, a diode in a level-shift circuit will also produce some attenuation due to the potential divider which is created by the a.c. resistance and the impedance of the current source device (Q2 in Figure 6.9). Digital circuits which use the level-shift circuit usually have sufficient noise margin so that the attenuation is not significant, and typically diodes with the same area as the MESFETs in the level-shift stage are used. However, in analog circuits the degradation of the stage gain below the ideal value of

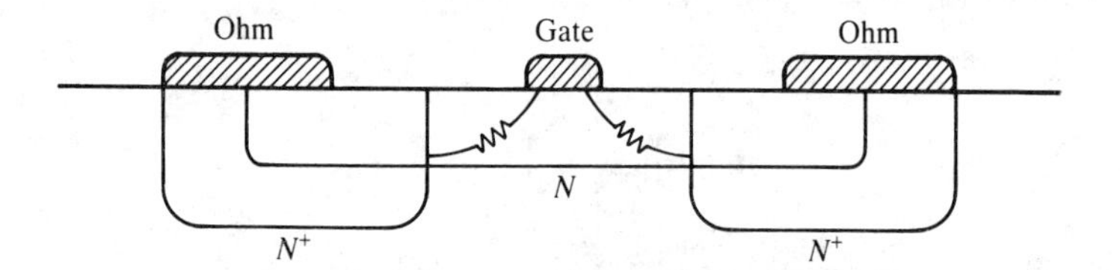

Figure 6.8 Parasitic resistance in Schottky diodes

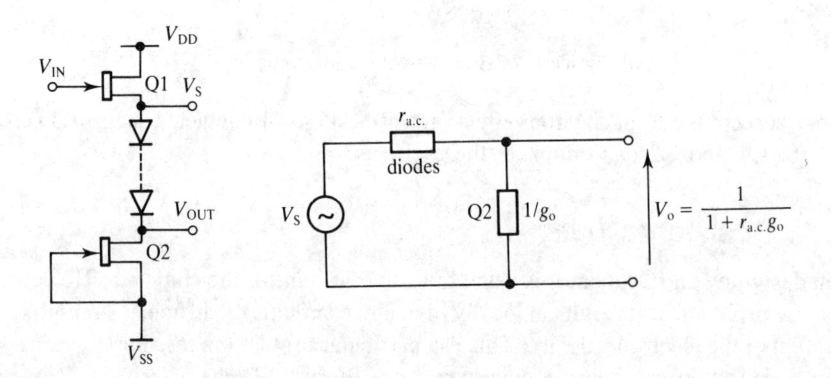

Figure 6.9 Attenuation in level-shift circuits

$(g_m/g_o)/[(g_m/g_o) + 2]$ may be unacceptable, and every attempt should be made to reduce the parasitic component of the a.c. resistance. This can be achieved by firstly reducing the physical separation between the Schottky and ohmic contacts, and secondly by increasing the perimeter as much as possible by adopting large aspect ratios for the Schottky contact. An example of a possible structure is shown in Figure 6.10.

6.2.9 Parasitic IGFETs

If a second-level of metallization is routed over resistor or MESFET devices, then a parasitic metal–insulator–semiconductor structure (i.e. an IGFET) will be created (Figure 6.11). In principle, signals on the metal track could modulate the resistivity of the semiconductor and thereby affect the circuit operation. In practice, this only happens to any significant extent with very large d.c. levels or fast signal changes; at low frequencies energy states

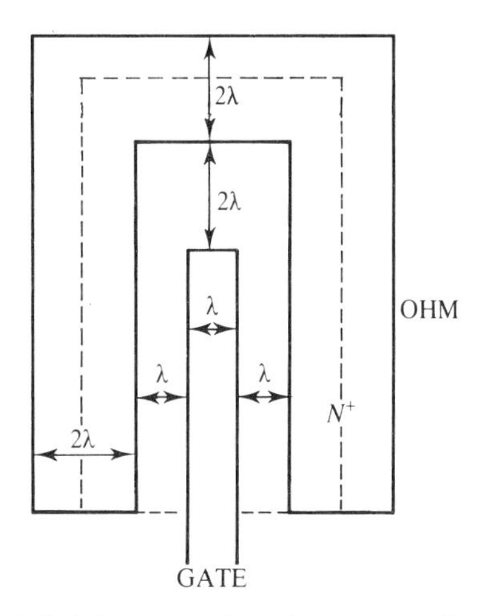

Figure 6.10 Minimizing parasitic resistance in a diode layout

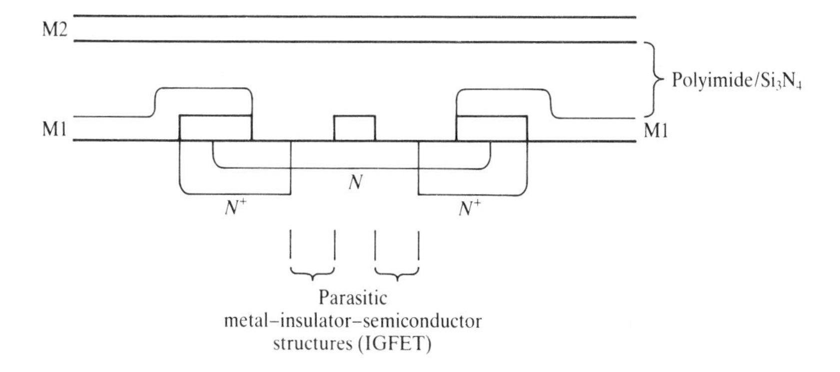

Figure 6.11 Parasitic IGFET structures

at the GaAs surface trap the charge induced by voltages on the metal track so that the semiconductor resistivity is not significantly affected. Power rails, for example, could thus be routed over the devices in a circuit to produce significant savings in the area required for the circuit layout. This is particularly advantageous in digital ICs where high packing densities are required for VLSI designs.

6.2.10 Compensating for process biasing

When features within an IC are fabricated there is usually a discrepancy between the 'drawn' width of the feature on the mask and the 'electrical' width which appears in the processed devices. This discrepancy is known as the process bias. It is generally manufacturer dependent and may occur for a variety of reasons. For example, implanted areas may diffuse sideways during the activation anneal step (hence the preference for flash annealing — see Chapter 5). Alternatively, the photoresist used in photolithographic etch or lift-off processes may be subject to some undercutting or shrinkage which results in the fabricated features being smaller or larger than the mask dimension. The effect of the process bias is to give fabricated devices which have different characteristics from the nominal values used in the initial design.

Similar differences between the fabricated and design characteristics will occur due to random and systematic variations associated with the different steps in the fabrication process. For example, any temperature gradient across the wafer during an activation anneal step may result in a corresponding variation in the threshold voltages across the wafer. To a reasonable approximation, this non-uniformity can be considered to produce effects equivalent to a variation in the width of the devices.

The process bias and the device non-uniformity become important if optimum circuit operation is dependent upon the ratio of device sizes within the circuit being maintained at a certain value. Digital circuits, for example, use the ratio of the widths of the switch and load devices to define the noise margin. If a DCFL logic gate requires a 6 : 1 ratio and is accordingly designed with a 30 μm wide switch and a 5 μm load, then a process bias of only +1 μm will change the width ratio to 5.17 : 1 which will result in severe reductions in the noise margin and circuit yield. Analog circuits may be similarly affected, for example, the input offset voltage in amplifiers and buffers is determined by the ratio of the device widths in the circuit. The IC should thus be designed so as to be tolerant to any process bias and device non-uniformity which may occur.

In principle, the process bias may be compensated during the mask-making process by increasing or decreasing the feature sizes accordingly. The difficulty associated with this is that the manufacturer's process bias figures will be an average, and the bias will not be the same on every process run nor uniform across a wafer. The device non-uniformity cannot be compensated for in this way at all. Circuits should thus be designed so that a degree of process tolerance is built in. Digital DCFL circuits which use a relatively low depletion-mode threshold voltage (e.g. −0.3 V) are thus preferred since the optimum width ratio is then close to unity (see Table 3.1) and relatively immune to different process bias effects on the switch and load devices. Likewise in analog positive level-shift circuits

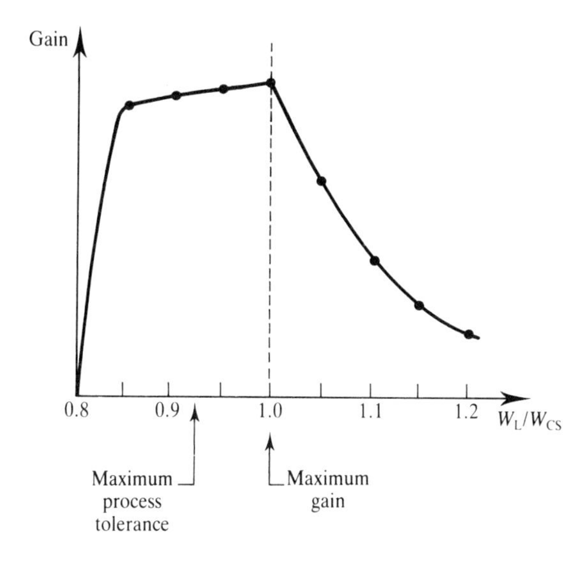

Figure 6.12 Differential amplifier gain vs load–current-source width ratio

(Section 4.2.3), it is generally better not to amalgamate the lower devices in the input buffer and level-shift stages so as to maintain the unity width ratio required by both stages.

Similarly, Figure 6.12 shows the dependence of the gain of a differential amplifier upon the load to current–source width ratio. At high and low values of this ratio the load and input devices, respectively, will not be in the saturation region of operation resulting in a loss of gain; it can be seen that the greatest process tolerance is obtained when the ratio is slightly less than unity. If the operation of a circuit depends upon a certain critical ratio being maintained, then 'Murphy's Law' can be relied upon to ensure that the circuit will never work at all!

6.2.11 Thermal gradients

The thermal conductivity of GaAs is approximately 60 percent of that of silicon, so any localized areas of high power dissipation will produce relatively severe temperature gradients. In circuits where the matching of component parameters is important (e.g. logic gate delays or amplifier characteristics), these temperature gradients can adversely affect the circuit operation and in amplifier designs thermal feedback may induce low frequency distortion [79].

High power dissipation areas will be produced in any areas with a relatively high packing density, or in any large width device such as used in an output buffer. Such local high packing densities should be eliminated if possible, and the output buffers in an IC should be distributed around the periphery so as to minimize the overall temperature gradient.

6.3 System considerations

Obtaining the maximum speed performance from an IC in a system requires design effort being expended on the IC environment, not just on the internal details of the IC. In the same way that the intrinsic performance of a transistor is degraded by parasitic components (e.g. source resistance and track capacitance in a GaAs IC), the performance of an IC can be severely degraded if the IC environment is not designed correctly. This section describes the prime considerations which must be dealt with when designing any high speed system, with particular reference to the requirements of GaAs as necessary.

6.3.1 Transmission line effects

At very high frequencies, the parasitic inductance present on long tracks cannot be neglected as in Section 6.2.2. In combination with the track parasitic capacitance, the parasitic inductance can cause severe distortion of signals, the most common aspect of which is 'ringing' on fast transitions (Figure 6.13). The distortion occurs due to interaction between the signal transmitted down the track and the reflection from the end, this interaction becoming significant in the situation where the track propagation time is comparable with the period of the signal or its harmonics. The cure for this distortion is based upon treating the track as a transmission line, and terminating it with a resistor having a value equal or close to the characteristic impedance. In this 'terminated' condition the track looks like an infinitely long line, and no reflection of the signal occurs at the termination. The signal distortion is then eliminated.

The characteristic impedance is determined by the track capacitance and inductance per unit length according to:

$$Z_o = \sqrt{\frac{L}{C}}$$

With a signal on an isolated track in free space this reduces to:

$$Z_o = \sqrt{\frac{\mu_o}{\epsilon_o}} = 367\ \Omega$$

If the signal return current flows in a parallel track, as shown in Figure 6.14(a), then the characteristic impedance is determined by [80]:

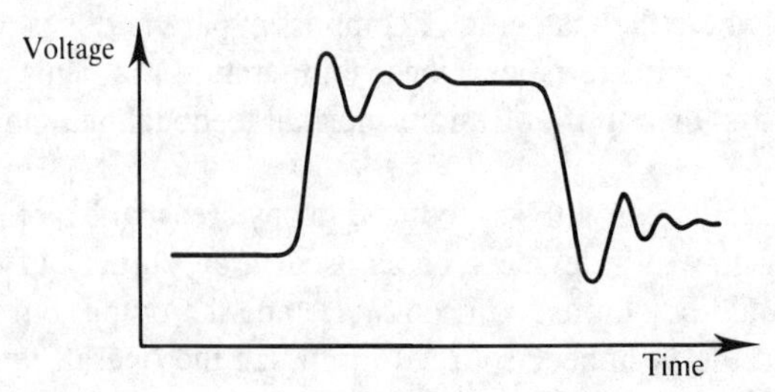

Figure 6.13 'Ringing' on fast signal transitions on long tracks

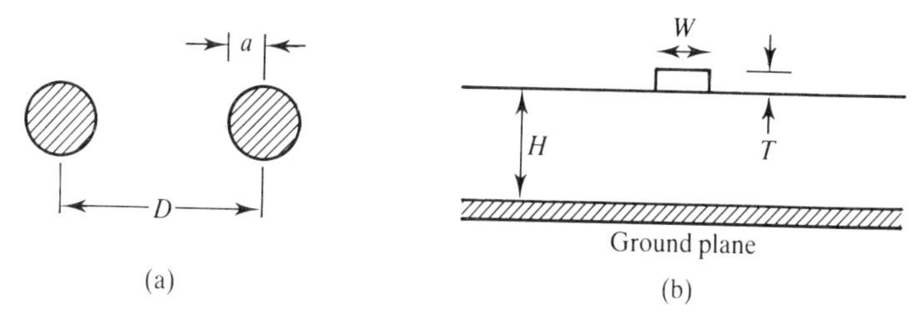

Figure 6.14 Dimensions used in characteristic impedance calculations for: (a) parallel tracks, and (b) track above a ground plane

$$Z_o = \frac{1}{\pi}\sqrt{\frac{\mu}{\epsilon}}\ln\frac{D}{a}$$

where the terms are as defined in the figure. More commonly the return current will flow in a ground plane as in Figure 6.14(b), the characteristic impedance is then [81]:

$$Z_o = \frac{87}{\sqrt{\epsilon_r + 1.41}}\ln\left(\frac{5.98H}{0.8W + T}\right)$$

As a rough guide, transmission line effects become important at frequencies where the signal wavelength is less than approximately 20 times the track length (i.e. $l > \lambda/20$) [82] (see Table 6.1). The track should be treated as a transmission line and terminated correctly if signals contain high frequency components with wavelengths which fall below this criterion. In digital systems, the maximum signal frequency is determined by the rise and fall times, not the maximum clock or data rate, and as a rough guide a bandwidth of three to five times the maximum signal rate should be assumed.

The wavelength of a signal depends on its velocity, which is itself dependent on the relative permittivity (ϵ_r) of the medium surrounding the track according to:

$$V_{ph} = \frac{c}{\sqrt{\epsilon_r}}$$

Table 6.1 Examples of track lengths above which lines should be treated as transmission lines

Track substrate	Average ϵ_r	Signal velocity (m s^{-1})	$\lambda/20$ (mm) at frequency:		
			500 MHz	1 GHz	2 GHz
p.c.b. ($\epsilon_r = 3$)	2	2.12×10^8	21.2	10.6	5.3
ceramic ($\epsilon_r = 10$)	5.5	1.28×10^8	12.8	6.4	3.2
GaAs ($\epsilon_r = 12$)	6.5	1.18×10^8	11.8	5.9	2.9

where $c = 3 \times 10^8$ m s^{-1}. In the case of a track running on the surface of a p.c.b., or on a GaAs or silicon IC, the value of ϵ_r to be used should be the average of the values for the materials above and below the track. As an example, in the case of a track running on the surface of a fibre glass p.c.b. ($\epsilon_r = 3$) with air ($\epsilon_r = 1$) above, then:

$$V_{ph} = \frac{c}{\sqrt{2}} = 2.12 \times 10^3 \text{ m s}^{-1}$$

The wavelength of a signal on this track is then determined by the standard relationship:

$$\lambda = \frac{V_{ph}}{f}$$

This wavelength should be compared with the length of the track to determine whether or not it should be treated as a transmission line. As examples, Table 6.1 lists values of ϵ_r, V_{ph} and $\lambda/20$ for various track substrates and frequencies. It should be clear from this table that transmission line effects can be safely ignored within an IC for all but the highest frequencies and line lengths. However, outside the chip where track lengths are much longer then these effects will become significant at relatively low frequencies.

6.3.2 Power supply decoupling

It was noted in Section 6.2.3 that reliable high speed operation of a circuit depends on the availability of a constant voltage, low impedance supply at the IC. This requirement is due to the fact that a high impedance supply will cause severe voltage fluctuations at the IC pins if changes in the current load occur during operation; these voltage fluctuations could cause malfunctions in the operation of the IC and would certainly degrade the available noise margin. The supply rails to the IC will have a finite resistance and inductance which can present a high impedance at high frequencies, and this can be aggravated by the characteristics of the power supply. Some means of reducing the supply impedance at the IC must thus be found. This is usually achieved by the addition of a decoupling capacitor connected between the supply rail and ground.

In connecting the decoupling capacitor, several considerations must be borne in mind. Firstly, the type of capacitor chosen is very important. Some types have a limited frequency response and can look inductive at very high frequencies! As general rules, tantalum capacitors provide better decoupling than electrolytic capacitors alone, and the best decoupling is provided by a pair of capacitors such as a large value component (usually polarized) to decouple long transients, in parallel with a small high frequency component (e.g. ceramic) to absorb fast transients on the supply. The combination of a 10 μF tantalum capacitor in parallel with a 10 nF ceramic component is often used, but beware! The exact characteristics of the combination will depend upon the characteristics of the supply and the parasitics present within the capacitor (particularly inductance); it is very easy to choose a pair which produce a parallel resonant circuit (and thus a very high impedance at some frequency) instead of a very low impedance. The supply rails should always be examined with a fast oscilloscope at the IC pins to ensure that a good, clean voltage supply is provided by the chosen decoupling capacitor [83].

The second and third considerations regarding the decoupling capacitor are closely related to the first. The capacitor should be connected as close to the IC pins as possible, and the leads should be as short as possible. This will then minimize the series inductance associated with the decoupling capacitor leads, and the supply impedance at the IC will be reduced as far as possible.

One further technique which may be used to safeguard the operation of the IC is the use of transient suppressors at the output of the power supply. This removes any large voltage spikes which may be induced by mains-borne interference.

6.3.3 Ground connections

One of the many factors limiting the speed of a system is the inductance associated with signal lines and other components, as this inductance prevents fast changes in current. In the case of the signal lines in a high speed system, the effects of the inductance associated with the line can be removed by treating the track as a transmission line and terminating with the characteristic impedance. However, there are many situations where the associated inductance cannot be removed, for instance, in connection with the terminating resistor itself or the power supply wiring. In these situations it is important that the inductance is at least minimized if maximum speed is to be achieved. One obvious solution is to minimize all wire lengths and also the length of the leads associated with components, but a second very important technique is to provide a 'ground plane' in the system.

A ground plane is what its name suggests: a continuous plane of grounded metal over which the system components are mounted. It is common practice to use a double or multi-layer p.c.b., one side is metallized as much as possible to provide the ground plane while signal and supply tracks are provided on the other layer(s). Components are usually mounted on the side with the ground plane; this allows very short lead lengths to ground, and also close spacing between components and the ground plane (Figure 6.15). This is especially important for components carrying high frequency currents, e.g. terminating resistors, ICs, transistors and decoupling capacitors.

The reduction of the system inductance by the ground plane relies upon the fact that the inductance of a loop is proportional to the loop area. Current from a signal source flowing in any track or component will have a return current flowing back to the ground connection of the signal source, and without a ground plane the current path may define a large loop area and thus a large inductance. With a ground plane though, the return current path can always flow directly under the track or component in question, and thus the loop area and inductance is small.

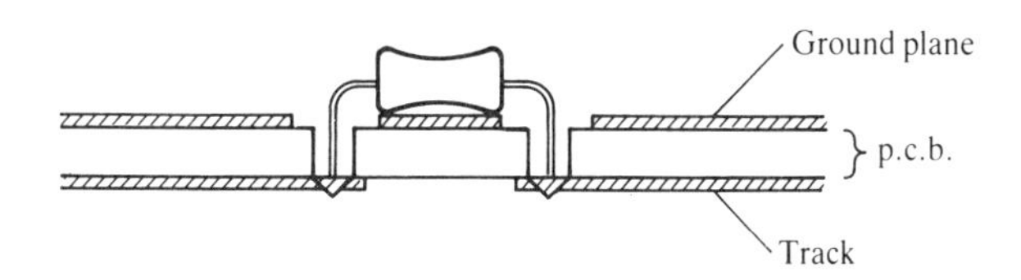

Figure 6.15 Component mounting on p.c.b.s with a ground plane

Although the ground plane also provides a low impedance connection to the ground rail, it must be remembered that the impedance is not zero and that potential differences may exist between various points on the plane. Some circuits (e.g. comparators) demand a common ground potential, and for these circuits the critical components should all be grounded at one point on the ground plane. In this case it may be necessary to have relatively long lead lengths, and thus to accept increased inductance in return for a common ground point.

7 □ Future developments

Future developments in GaAs ICs may be centred on the use of new device structures now being investigated. Chiefly these investigations concern the high electron mobility transistor (HEMT) and the heterojunction bipolar transistor (HJBT), but effort is also being directed towards the integration of optical and electronic functions on one chip. These developments are described in the following sections.

7.1 High electron mobility transistors (HEMTs)

The advantage of GaAs transistors over silicon devices depends chiefly upon the higher electron mobility in GaAs. This higher mobility allows GaAs transistors to be fabricated with a lower resistance, and thus a higher speed, than comparable silicon transistors.

If maximum speed is required, the intuitive solution is to use a high doping level so as to minimize the resistances in the transistor. This solution, however, has limited usefulness for several reasons: firstly, the capacitances in the circuit will increase due to a reduction in the depletion layer thicknesses as the doping increases; secondly, the electron mobility decreases as the doping level increases (due to increased scattering by the dopants); and thirdly, there are practical limitations to the maximum doping level which can be used since extremely high levels (typically above 10^{20} cm^{-3}) will introduce a large number of dislocations and defects into the crystal lattice. If maximum speed is really required, some means must be found of producing high carrier concentrations in undoped material so as to maximize the electron mobility. At first, these requirements might seem to be contradictory, but they are achieved in the device structure used by the HEMT. Electron mobilities typically of the order of 50,000 cm^2 V^{-1} s^{-1} at 77 K and at 300 K, respectively can be achieved; these values should be compared with the typical values of 4,000 cm^2 V^{-1} s^{-1} and 800 cm^2 V^{-1} s^{-1} exhibited by doped GaAs and silicon at 300 K.

In the HEMT structure (shown in Figure 7.1), a highly doped layer of $Al_{(1-x)}Ga_xAs$

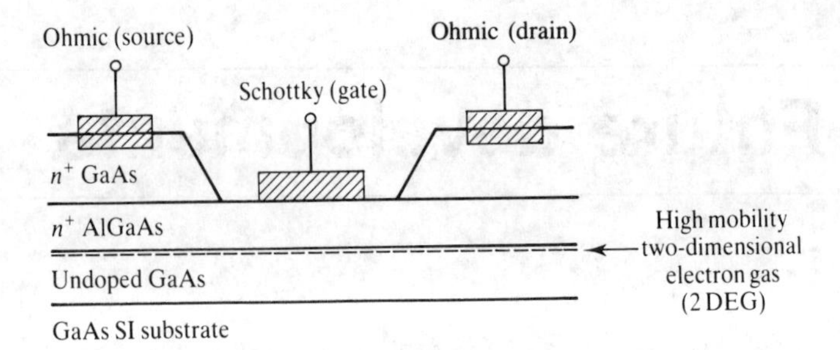

Figure 7.1 A HEMT structure

is grown on top of a high purity, undoped, layer of GaAs. The ternary compound $Al_{(1-x)}Ga_xAs$ is formed from a mixture of GaAs and AlAs in the ratio of x to $(1 - x)$; x is typically around 0.3. The band structure of the interface between these two layers is such that electrons from the doped $Al_{(1-x)}Ga_xAs$ layer will diffuse into the undoped GaAs to sit in a very thin layer at the interface. In the ideal device this layer actually has no thickness, and so is referred to as a 'two-dimensional electron gas' (2DEG), although real devices approach this ideal with variable degrees of success. Provided the undoped layer is very pure, and the interface free from defects and strain, the mobility of the transferred electrons will be very high; improvements of the order of 2 : 1 at room temperature (300 K) and 10 : 1 at 77 K are typical compared to doped GaAs.

The conductivity of the high mobility electron layer can be modulated by controlling the number of carriers within the 2DEG. This can be achieved by varying the width of a depletion region which extends into the 2DEG from the doped $Al_{(1-x)}Ga_xAs$ layer. The depletion region is created by a Schottky gate contact on the $Al_{(1-x)}Ga_xAs$ layer. The device characteristics are analogous to that of the standard GaAs MESFET, and enhancement or depletion-mode devices can be created simply by suitably selecting the thickness of the $Al_{(1-x)}Ga_xAs$ layer so that the 2DEG is either depleted or full at zero gate bias. The Schottky gate contact to $Al_{(1-x)}Ga_xAs$ has a higher built-in voltage than a similar contact to GaAs. Thus this HEMT structure has the additional advantage of allowing a larger voltage swing before forward gate conduction occurs; this results in a better noise margin than in GaAs MESFET-based circuits.

Ohmic contacts to the HEMT are usually made via a doped layer of GaAs grown on top of the $Al_{(1-x)}Ga_xAs$ layer. This layer gives a lower (parasitic) contact resistance than would contacts made directly to $Al_{(1-x)}Ga_xAs$ due to the smaller bandgap of GaAs; the wide bandgap of $Al_{(1-x)}Ga_xAs$ makes it difficult to achieve low resistivity ohmic contacts.

To date HEMT structures have been used to fabricate relatively complex gate array and static RAM ICs [84, 85], with gate delays of the order of 25 ps per gate at 77 K. The main difficulty preventing the widespread application of HEMT technology is the complex processing techniques required to produce good structures; these depend on MOCVD or MBE growth techniques which are not suited for high volume IC production at present.

7.2 Heterojunction bipolar transistors (HJBTs)

The gain of a bipolar transistor is dependent upon the ratio of the collector to emitter current, I_C/I_E, and usually this ratio (α) is required to be as close to unity as possible. The value of α is dependent upon the ratio of electron to hole currents through the base-emitter junction (β_{max}) [86], which is dependent upon the value of the expression:

$$\frac{N_E}{N_B W_B} \exp\left(\frac{\Delta E_g}{kT}\right)$$

N_E and N_B are the dopant concentrations in the emitter and base, respectively; W_B is the base width; and ΔE_g is the difference between the bandgaps in the emitter and base regions. To obtain values of α close to unity in a conventional homojunction device (i.e. a transistor fabricated from p–n junctions in one semiconductor material), the emitter must be more highly doped than the base; in this condition ΔE_g will be slightly negative since the bandgap reduces as the doping level increases.

Chapter 2 included a discussion of the properties of GaAs homojunction bipolar transistors. If we restrict our attention to n–p–n devices (which have a better frequency performance than p–n–p devices), it was stated that GaAs bipolar transistors do not offer any significant speed advantage over silicon transistors due to a relatively low hole mobility and thus a correspondingly high parasitic base resistance. However, this can be overcome if an alternative structure is fabricated using an $Al_{(1-x)}Ga_xAs$ emitter (where x is typically around 0.3); this structure is known as a heterojunction bipolar transistor [87]. The bandgap of $Al_{(1-x)}Ga_xAs$ is greater than GaAs (by an amount dependent upon the value of x), so ΔE_g becomes positive and the exponential term in the equation above can be traded off against the ($N_E/N_B W_B$) term while maintaining α close to unity. In other words, the use of a wide bandgap emitter allows the base doping level to be increased without degrading the transistor gain. This increased base doping allows a much reduced parasitic base resistance, and thus an improved frequency response. The base-emitter capacitance, which also determines this frequency response, can be kept small by making the emitter region less heavily doped than the base region. Calculations suggest that AlGaAs/GaAs HJBTs can be fabricated with an upper useful frequency limit of the order of 100–200 GHz, and experimental values of the order of 100 GHz [88] have been achieved. The corresponding value for GaAs MESFETs is of the order of 20 GHz.

One possible structure for a AlGaAs/GaAs HJBT is shown in Figure 7.2. As with the HEMT structure, this has a GaAs layer grown on top of the AlGaAs emitter to reduce the parasitic ohmic contact resistance. Possible variant structures are based around using a 'collector-up' structure (Figure 7.3(a)), using a Schottky contact for the collector (Figure 7.3(b)) and also using a wide bandgap collector (the so-called double heterojunction bipolar transistor, or DHJBT) (Figure 7.3(c)). The main disadvantages of the HJBT are associated with the complex processing required (again dependent on MOCVD or MBE growth techniques) and the non-planar structure which can give step-coverage problems for interconnecting tracks. In compensation, the drive capability of these transistors is much superior to that of GaAs MESFETs and this reduces the dependency of the propagation

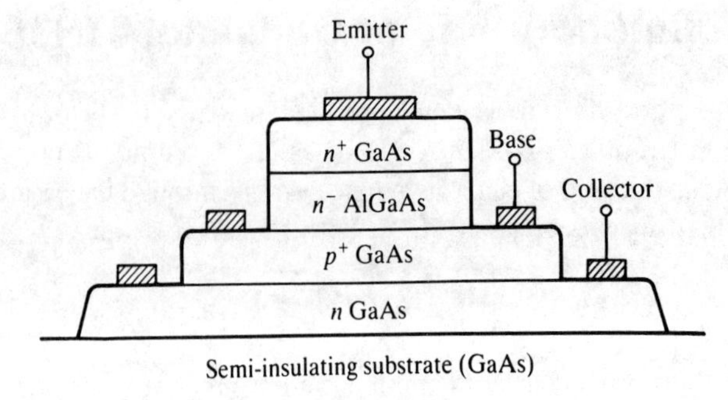

Figure 7.2 A HJBT structure

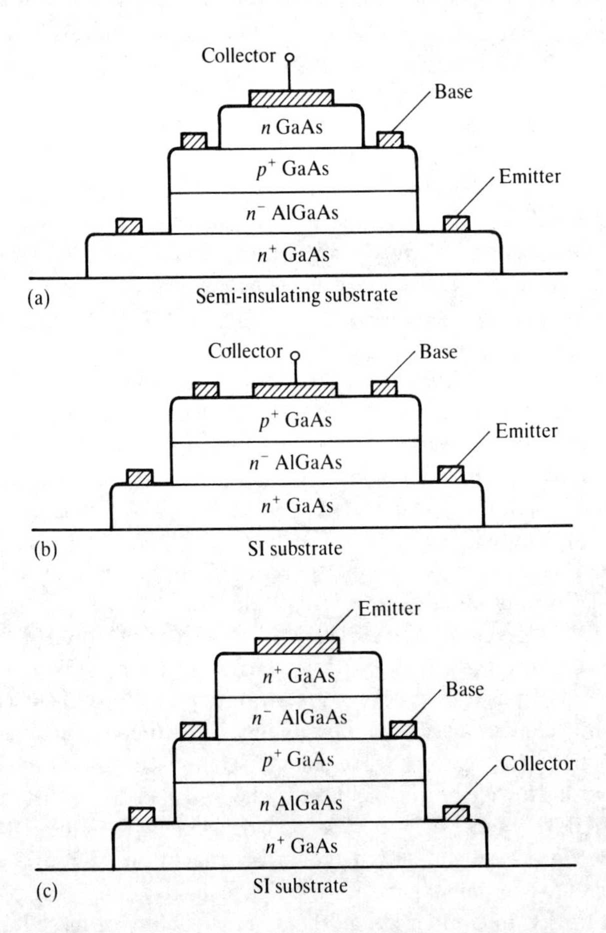

Figure 7.3 Alternative structures for HJBTs: (a) 'collector-up'; (b) Schottky collector; (c) double heterojunction bipolar transistor

delay on loading. Furthermore, the reproducibility of logic gates built using these devices is also much improved since the device characteristics are dependent upon the logarithm of the doping levels, and are thus less sensitive to processing variations than are MESFETs where the threshold voltage is directly proportional to the doping level. These properties make AlGaAs/GaAs HJBTs ideal for very high-speed VLSI circuits [89–91]; predictions of 10 ps gate delays at room temperature have been made, and digital ICs operating at 20 Gbit s^{-1} have been fabricated [92]. In the design of circuits using HJBTs, advantage can also be taken by making use of design techniques developed for silicon bipolar transistor circuits. This contrasts favourably with the situation for MESFET-based circuits where new design techniques have had to be developed.

7.3 Opto-electronic integration

As system clock and data rate requirements increase to very high speeds, one of the main technical obstacles to be overcome besides that of increasing the intrinsic device speed is that of maintaining this speed when the devices are connected together, particularly when these devices lie on separate ICs. As the device dimensions are shrunk so as to obtain higher operating speeds, so the effects of the parasitic components become increasingly more significant. The operating speed of many state-of-the-art ICs is limited by these parasitics rather than by the intrinsic operating device speed. At very high speeds the limiting parasitics are often associated with the I/O interfaces of the circuit. The parasitics will be higher here than within the internal circuitry due to the additional capacitance presented by the device packaging and the additional inductance presented by bondwires, and also due to the fact that external connections are generally longer and wider than the internal connections.

The main problems imposed by the I/O parasitics are concerned with power dissipation and cross-talk. There will generally be a capacitive load present on each output from an IC which is due to the parasitics associated with the packaging, the interconnecting track and the inputs of the driven ICs. This can be reduced by treating the track as a transmission line and terminating it accordingly (see Chapter 6), but some stray capacitance will usually remain. Ensuring that the driven inputs see the required voltage swing at increasing speeds means that increasing levels of current will be required from the output buffers since

$$I = C\frac{\delta V}{\delta t}$$

Thus, as data rates increase so the power dissipation associated with the output buffers also increases. The total power budget for the IC may thus set an operating speed limit which is well below that expected from the intrinsic device speed, because of the high power dissipation in the buffers. The problem of cross-talk is also due to the parasitic capacitance in the system; as data rates increase, the amount of interaction which occurs between signals via this capacitance will increase accordingly.

One possible method of alleviating these problems is to replace electrical chip-to-chip communicating signals with optical signals. Transmitting signals over optical fibers will eliminate any cross-talk between the signals in the fibers. However, if the optical transmitters and receivers are distinct from the electronic components then there will be little change in the parasitic capacitance associated with the packaging and the inputs compared to an all-electronic system, and the problems of high power dissipation and cross-talk between the electrical signals will remain. This situation changes, however, if the optical and electronic components can be integrated together, as it will then be possible to greatly reduce the parasitic capacitance load, and hence to greatly reduce the associated problems.

The lattice mismatch between silicon and suitable light-emitting materials (e.g. AlGaAs) sets up a strain in the device structures which makes the manufacture of good optical devices integrated with electronic components extremely difficult. In contrast, GaAs and $Al_{(1-x)}Ga_xAs$ have very similar lattice parameters, and this makes the integration of opto-electronic systems relatively easy to achieve.

Integration of optical and electronic components is required at two interfaces: the electronic–optical interface for signals being transmitted from the IC, and the optical–electronic interface for signals being received. At the electronic–optical interface transmission is usually carried out using laser structures fabricated on the electronic substrate (see Figure 7.4(a)). Although this technique overcomes the limitations imposed by parasitics at the output, the optical outputs suffer from the same problem of high localized power dissipation as do electronic outputs: lasers require relatively high powers for efficient operation and the task of dissipating this power may again impose a limit to the number of high-speed transmitters which can be present on an IC. An alternative method of

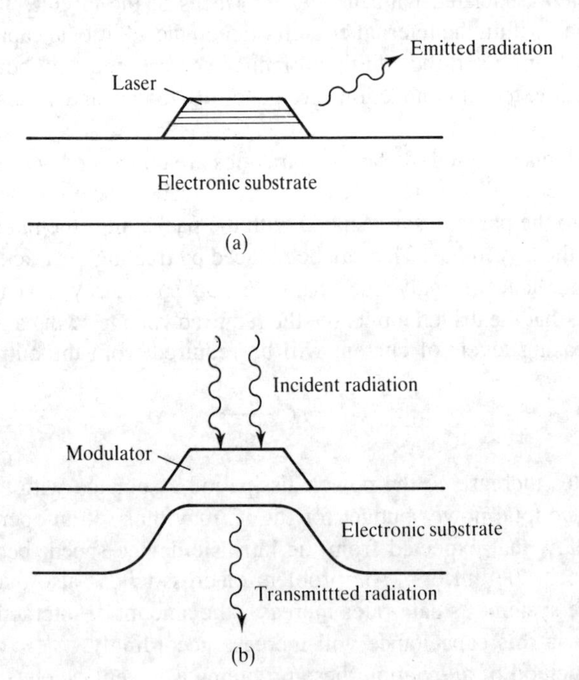

Figure 7.4 The electronic–optical interface: (a) lasers; (b) modulators

overcoming this limitation involves using an external laser source, controlling the intensity of the light transmitted through (or reflected from) the IC using **modulator** devices as shown in Figure 7.4(b).

A schematic of a transmission modulator is shown in Figure 7.5 [93]. A multi-quantum well (MQW) structure is built up from alternating very thin layers of GaAs and AlGaAs (or InP/InGaAs). The energy band structure within the MQW can be controlled by varying the electric field across it, and with appropriate selection of the MQW layer thicknesses the absorption at any selected optical frequency can be made a function of the applied bias as shown in Figure 7.6 [94]. This method suffers from a relatively low on/off contrast ratio at the low voltages commonly present in ICs. However, the problem of high power dissipation is overcome and future devices show promise of an improved contrast ratio suitable for use in complex systems [95].

At the optical–electronic interface, optical radiation is converted into electrical signals. Two types of transducer are most commonly used at this interface: the avalanche photodiode (APD), and the PIN photodiode. Both structures rely on the optical generation of hole–electron pairs in a diode structure to produce a signal current. In the APD structure the generated carriers are accelerated under a high electric field so that avalanche breakdown occurs, and these detectors thus give the advantage of having a degree of built-in amplification. Unfortunately, relatively high biases (typically 60 V) are required to achieve avalanche breakdown, and these biases are not generally suited to use in ICs. The alternative PIN diode is thus the most widely used, although their use necessitates the availability of very low noise amplifiers to amplify the detected signal.

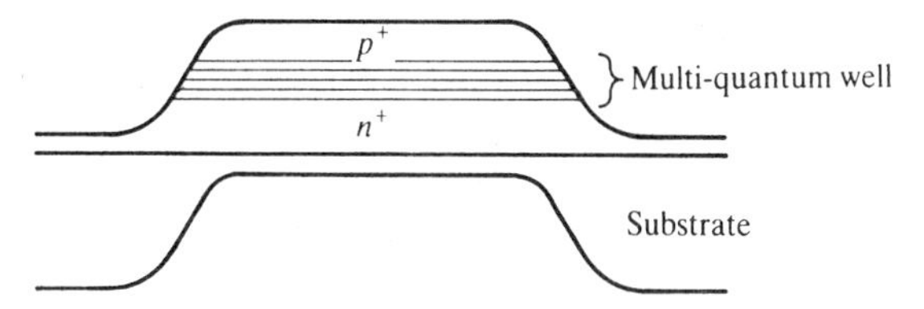

Figure 7.5 A transmission modulator structure

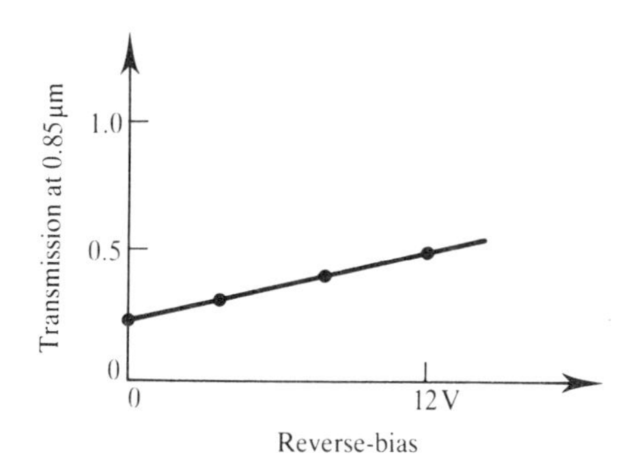

Figure 7.6 Variation of optical transmission with reverse-bias in a modulator

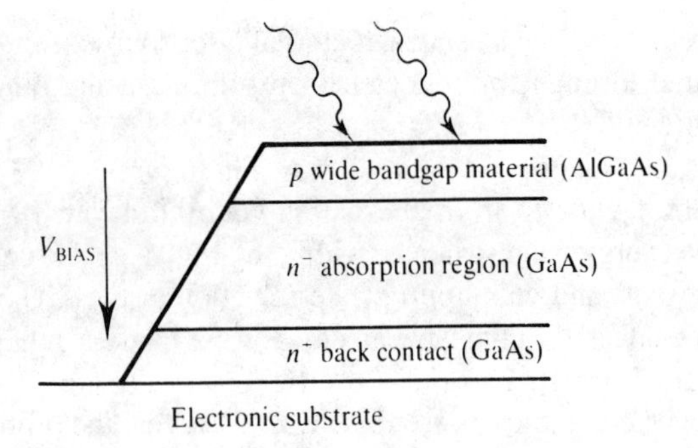

Figure 7.7 A PIN photodiode structure

The structure of a PIN diode is shown in Figure 7.7. Light is absorbed in the depletion region of a reverse-biased $p-n$ junction; this junction is very lightly doped in order to make the depletion region as thick as possible and thus maximize the absorption within it. (The PIN notation for this detector derives from this *P*-*I*ntrinsic-*N*-type structure.) The upper layer is often a wide bandgap material to minimize the absorption in the top contact layer. Typically semiconductor systems such as AlGaAs/GaAs or InGaAs/InP are used. Fabrication of these detectors within an opto-electronic IC (OEIC) is difficult since the optimum layer doping densities and thicknesses will be different for the optical and electronic components, and the growth of the optical layers may affect the properties of the transistors undesirably. However, despite these difficulties, excellent sensitivities have been achieved at multi-gigabit bit rates [96]. At low frequencies better results can be achieved using separate detectors and amplifiers, as their performance can then be individually optimized. However, as bit rates increase so the effect of the parasitics on the interconnect between the optical and electronic components become increasingly important, and these parasitics will limit the ultimate performance of systems based upon separate optical and electronic components. If data rates are to continue to increase in the future, then the availability of fully integrated opto-electronic systems will become an essential requirement of future systems, and GaAs IC technology will play a key role in their development.

References

1. Mayer, D. C., A. Lee, A. R. Kramer and K. D. Miller, 'High performance CMOS/SOS circuits in SPEAR material', *IEEE J. Solid State Circuits*, vol. 25, 1990, pp. 318–21.
2. Zuleeg, R., 'Radiation effects in GaAs FET devices', *Proc. IEEE*, 77, 1989, pp. 389–407.
3. Zuleeg, R., J. K. Notthoff and G. L. Troeger, 'Channel and substrate currents in GaAs FETs due to ionising radiation', *IEEE Trans. Nucl. Sci.*, NS-30, 1983, pp. 4151–6.
4. Redman-White, W., R. Dunn, R. Lucas and P. Smithers, 'Radiation-hard AGC stabilised SOS crystal oscillator', *IEEE Trans. Electron. Devices*, Ed-38, 1990, pp. 1730–6.
5. Roosild, S. A., 'DARPA plans and pilot production line project', *IEEE Conference on Computer Design*, Port Chester, New York, NY, October 1984, Silver Spring, MD: IEEE Comput. Society Press, 1984.
6. Zuleeg, R., J. K. Northoff and G. L. Troeger, 'Double ion-implanted complementary GaAs JFETs', *IEEE Electron. Device Lett.*, EDL-5, 1984, pp. 21–3.
7. Graffeuil, J., K. Tantrarongroj and J. F. Sautereau, 'Low frequency noise physical analysis for the improvement of the spectral purity of GaAs FET oscillators', *Solid State Electron.*, 25, 1982, pp. 367–74.
8. Tsironis, C., J. Graffenil, F. Henze and Z. Hadjoub, 'Low frequency noise in GaAs MESFETs', *GaAs and Related Compounds Symposium*, Biarritz, 26–8 September 1984, Bristol: Institute of Physics, 1984.
9. Wilmsen, C. W. (ed.), *Physics and Chemistry of III–V Compound Semiconductor Interfaces*, New York, NY: Plenum, 1985.
10. Lee, W. S. and J. G. Swanson, 'Switching behaviour of Al_2O_3–nGaAs MESFETs', *Electron. Lett.*, 18, 1982, pp. 1049–51.
11. Laude, D. and G. Noufer, 'Design considerations for commercial GaAs digital ICs', *IEEE Custom Integrated Circuits Conference*, Portland, OR, 20–3 May 1985, New York: IEEE, 1985, pp. 421–5.
12. Steiner, K., H. Mikami, Y. Kitaura and N. Uchitomi, 'Minimum size effects in asymmetric tilt-angle-implanted LDD–WN_x GaAs MESFETs', *IEEE Trans. Electron. Devices*, ED-38, 1991, pp. 1730–6.
13. Wan, C. F., H. Shichijo, R. D. Hudgens, D. L. Plumton and L. T. Tran, 'A comparison study of GaAs E/D MESFETs fabricated with self-aligned and non-self-aligned processes', *IEEE GaAs IC Symposium Technical Digest*, Portland, OR, 13–16 October 1987, New York: IEEE, 1987, p. 133.

14. Camacho-Penalosa, C. and C. S. Aitchison, 'Modelling frequency dependence of output impedance of microwave MESFETs at low frequency', *Electron. Lett.*, 21, 1985, pp. 528–9.
15. Kawasaki, H. and J. Kasahara, 'Low frequency dispersion of transconductance in GaAs JFETs and MESFETs with an ion-implanted channel layer', *IEEE Trans. Electron. Devices*, ED-37, 1990, pp. 1789–95.
16. Tan, K. L., 'A self-aligned gate GaAs MESFET technology for low-power sub-nanosecond static RAM fabrication', *IEEE GaAs IC Symposium Technical Digest*, Portland, OR, 13–16 October 1987, New York: IEEE, p. 121.
17. Matsunaga, N., M. Miyazaki, Y. Umemoto, J. Shigata, H. Tanaka and H. Yamazawa, 'GaAs MESFET technologies with 0.7 micron gate length for 4kb 1ns static RAM', *IEEE GaAs IC Symposium Technical Digest*, Portland, OR, 13–16 October 1987, New York: IEEE, p. 129.
18. Nagel, L., 'SPICE2: A Computer Program to Simulate Semiconductor Circuits', Electronics Research Laboratory, University of California, Berkeley, CA, Memo ERL-M592, 1976.
19. White, W. A. and M. R. Namordi, 'GaAs MESFET model adds life to SPICE', *Microwaves RF*, MTT-28, September 1984, p. 200L-P.
20. Curtice, W. R., 'A MESFET model for use in the design of GaAs integrated circuits', *IEEE Trans. Microwave Theory Techniques*, MTT-28, 1980, pp. 448–56.
21. Jastrzebski, A. K., 'MESFET modelling and parameter extraction', *GaAs Technology and Its Impact on Circuits and Systems*, D. G. Haigh and J. Everard (eds), London: Peter Peregrinus, 1989, ch. 3.
22. Toumazou, C. and D. G. Haigh, 'Design of a high-gain, single-stage operational amplifier for GaAs switched-capacitor filters', *Electron. Lett.*, 23, 1987, pp. 752–4.
23. '16G020 programmable single-gate GaAs D-MESFET array', *GaAs IC Data and Designers Guide*, Newbury Park, CA: Gigabit Logic Inc., 1988.
24. Cirillol, N. C., D. K. Arch, P. J. Vold, B. K. Betz, I. R. Mactaggart and B. L. Grung, '8×8 bit pipelined parallel multiplier utilising self-aligned gate n^+(AlGa)As/GaAs MODFET IC technology', *IEEE GaAs IC Symposium Technical Digest*, Portland, OR, 13–16 October 1987, New York: IEEE, 1987, p. 257.
25. Lee, C. P., R. Zucca and B. M. Welch, 'Saturated resistor loads for GaAs ICs', *IEEE Trans. Electron. Devices*, ED-29, 1982, pp. 1103–9.
26. Lundgren, R. E., C. F. Krumm and R. F. Lohr, 'Enhancement-mode GaAs MESFET logic', *GaAs IC Symposium*, Lake Tahoe, CA, September 1979, New York: IEEE, 1979.
27. Van Tuyl, R. and C. Liechti, 'High-speed integrated logic with GaAs MESFETs', *IEEE J. Solid State Circuits*, SC-9, 1974, pp. 269–76.
28. Eden, R. C., B. M. Welch and R. Zucca, 'Low power GaAs digital ICs using Schottky diode–FET logic', *IEEE International Solid State Circuits Conference*, San Francisco, CA, 15–17 February 1978, New York: IEEE, 1978, pp. 68–9.
29. Katsu, S., S. Nambu, A. Shimano and G. Kano, 'A source-coupled FET logic — a new current-mode approach to GaAs logics', *IEEE Trans. Electron. Devices*, ED-32, 1985, pp. 1114–8.
30. Mellor, P. J. T. and A. W. Livingstone, 'Capacitor-coupled logic using GaAs depletion-mode FETs', *Electron. Lett.*, 16, 1980, p. 749.
31. Hashizume, N., H. Yamada, T. Kojima and K. Matsumoto, 'Low-power-gigabit logic by GaAs SSFL', *Electron. Lett.*, 17, 1981, p. 553.
32. Livingstone, A. W. and D. Welbourn, 'Design considerations of coupling capacitors in GaAs integrated circuits', *Japan. J. Appl. Phys.*, 22, Suppl. 22-1, 1983, pp. 393–6.
33. Eden, R. C., 'Capacitor diode FET logic (CDFL) circuit approach for GaAs D-MESFET ICs', *IEEE GaAs IC Symposium Technical Digest*, Boston, MA, 23–5 October 1984, New York: IEEE, 1984, pp. 11–14.

34. Tanaka, K., H. Nakamura, Y. Kawakami, M. Akiyama, T. Ishida and K. Kaminishi, 'Super buffer FET logic (SBFL) for GaAs gate arrays', *IEEE Custom Integrated Circuits Conference*, Portland, OR, 20–3 May 1985, New York: IEEE, 1985, pp. 425–8.
35. Topping, J., *Errors of Observation and their Treatment*, London: Chapman & Hall, 1972.
36. Green, D. H., *Modern Logic Design*, Wokingham: Addison-Wesley, 1986.
37. Dierling, C. E., 'Interfacing Picologic and Nanoram GaAs ICs to other logic families', *GaAs IC Data Book*, Newbury Park, CA: Gigabit Logic Inc., 1988.
38. Canfield, P., J. Medinger and L. Forbes, 'Buried channel GaAs MESFETs with frequency independent output conductance', *IEEE Electron. Devices Lett.*, March 1987, pp. 88–9.
39. Haigh, D. G., C. Toumazou and A. K. Betts, 'Switched capacitor circuits and operational amplifiers', *GaAs Technology and Its Impact on Circuits and Systems*, D. G. Haigh and J. Everard (eds), London: Peter Peregrinus, 1989, ch. 10.
40. Larson, L. E., K. W. Martin and G. C. Temes, 'GaAs switched capacitor circuits for high speed signal processing', *IEEE J. Solid State Circuits*, SC-22, 1987, pp. 971–81.
41. Harrold, S. J., 'Switch-driver suitable for high-order switched-capacitor filters implemented in GaAs', *Electron. Lett.*, 24, 1988, pp. 982–4.
42. Phillips, J. A., S. J. Harrold and G. K. Barker, 'GaAs sampled-analogue integrated circuits', *J. IERE*, 51 (Suppl.), 1987, pp. S35–43.
43. Toumazou, C. and D. G. Haigh, 'Level-shifting differential to single-ended converter circuits for GaAs MESFET implementations', *Electron. Lett.*, 23, 1987, pp. 1053–5.
44. Allan, P. E. and G. M. Breevoort, 'An analogue circuit perspective of GaAs technology', *Proceedings of IEEE International Symposium on Circuits and Systems*, Philadelphia, PA, 4–7 May 1987, New York: IEEE, 1987.
45. Haigh, D. G., C. Toumazou, S. J. Harrold, K. Steptoe, J. I. Sewell and R. Bayruns, 'Design optimisation and testing of a GaAs switched-capacitor filter', *IEEE Trans. Circuits Systems*, 38, 1991, pp. 825–37.
46. Lindquist, P. F., 'Model of related electrical properties and impurity concentrations in semi-insulating GaAs', *J. Appl. Phys.*, 48, 1977, p. 1262.
47. Kaminska, M., J. Lagowski, J. Parsey, K. Wada and H. C. Gatos, 'Oxygen induced levels in GaAs', *International Symposium on GaAs and Related Compounds*, Oiso, Japan, 20–3 September 1981, Bristol: Institute of Physics, pp. 197–202.
48. Mortensen, P., 'Gallium arsenide in Japan', *Electron. Power*, 31, 1985, pp. 115–18.
49. Matsuura, H., H. Nakamura, Y. Sano, T. Egawa, T. Ishida and K. Kamimishi, *IEEE GaAs IC Symposium Technical Digest*, Monterey, CA, 12–14 November 1985, New York: IEEE, 1985, p. 167.
50. Miyazawa, S., Y. Ishii, S. Ishida and R. Nanishi, 'Direct observation of dislocation effects on threshold voltage of GaAs field effect transistors', *Appl. Phys. Lett.*, 43, 1983, pp. 853–5.
51. Winston, H. V., A. T. Hunter, H. M. Olsen, R. P. Bryan and R. E. Lee, 'FET arrays on In-alloyed GaAs substrates', *Proceedings of Semi-Insulating III–V Materials Conference*, Kah-nee-ta, OR, 24–6 April 1984, Nantwich: Shiva Publishing, 1984, pp. 402–5.
52. Thomas, R. N., S. McGuigan, G. W. Eldridge and D. L. Barrett, 'Status of device qualified GaAs substrate technology for GaAs ICs', *Proc. IEEE*, 76(7), 1988, pp. 778–91.
53. Tersashima, K., K. Katsumata, F. Orito, T. Kitna and T. Fukuda, 'Electrical resistivity of undoped GaAs single crystals grown by magnetic field LEC technique', *Japan. J. Appl. Phys.*, 22, 1983, pp. L325–7.
54. Jacob, C., M. Duseaux, J. P. Farges, M. M. B. van den Boom and P. J. Roksnoer, 'Dislocation-free GaAs and InP crystals by isoelectronic doping', *J. Cryst. Growth*, 61, 1983, pp. 417–24.
55. Hobgood, H. M., R. N. Thomas, D. L. Barrett, G. W. Eldridge, M. M. Sopira and M. C.

Driver, 'Large diameter, low dislocation In-doped GaAs: grwoth, characterization and implications for FET fabrication', *Proceedings of Semi-Insulating III–V Materials Conference*, Kah-nee-ta, OR, 24–6 April 1984, Nantwich: Shiva Publishing, 1984, pp. 149–56.
56. Shaw, D. W., 'Kinetics of transport and epitaxial growth of GaAs with GaAs–$AsCl_3$ system', *J. Cryst. Growth*, 8, 1971, p. 117.
57. Olson, G. H. and T. Z. Zamerowski, 'Crystal growth and properties of binary ternary and quarternarys (In, Ga) (As_3P) alloys grown by the hydride vapour phase epitaxy technique', *Prog. Cryst. Growth Characterisation*, 2, 1979, p. 309.
58. Miers, T. H., 'The influence of source stability on the purity and morphology of GaAs grown in the Ba/$AsCl_3$/H_2 system', *Institute of Physics Conference Series*, 65, Bristol: Institute of Physics, 1983, pp. 125–32.
59. Cho, A. Y. and J. R. Arthur, 'Molecular beam epitaxy', *Prog. Solid State Chem.*, 10, 1975, p. 157.
60. Lindhard, J., M. L. Scharff and H. E. Schiott, 'Range concepts and heavy ion ranges. Notes on atomic collisions II', *Matematisk-Fysike Meddelelser Konglige Danske Videnskabermes Selskab*, 33, 1963.
61. Kotera, N., K. Yamashita, Y. Hatta, T. Kimoshita, M. Miyazaki and M. Maeda, 'Laser drivers and receiver amplifiers for 2.4 Gbit/s optical transmission WSi-gate GaAs MESFETs', *IEEE GaAs IC Symposium Technical Digest*, Portland, OR, 13–16 October 1987, New York: IEEE, 1987, p. 103.
62. Ransome, S. J., L. G. Hipwood and M. J. Wheeler, 'Surface patterning of GaAs wafers for monolithic integrated circuits', *GEC J. Res.*, 4, 1986, pp. 91–103.
63. Hatzakis, M., B. J. Canavello and J. McShaw, 'Single step optical lift-off process', *IBM J. Res. Develop.*, 24, 1980, p. 452.
64. Bartle, D. C., D. C. Andrews, J. D. Grange, P. G. Harris, A. D. Trigg and D. K. Wickenden, 'Plasma enhanced deposition of silicon nitride for use as an encapsulant for silicon ion-implanted gallium arsenide', *Vacuum*, 34, 1984, pp. 315–20.
65. Anderson, C. L., K. V. Vaidyanathan, H. L. Dunlap and G. S. Kamath, 'Capless annealing of ion-implanted GaAs by a melt-controlled ambient technique', *J. Electrochem. Soc.*, 127, 1980, pp. 925–7.
66. Tabatabaie-Alavi, K., A. N. M. Masum Choudhury, C. G. Fonstad and J. C. Gelpi, 'Rapid thermal annealing of Be, Si and Zn implanted GaAs using an ultra high power argon arc lamp', *Appl. Phys. Lett.*, 43, 1983, pp. 505–7.
67. Yokoyama, N., T. Ohnishi, K. Odaui, H. Onodera and M. Abe, 'TiW silicide gate self-alignment technology for ultra high speed GaAs MESFET LSI/VLSI', *IEEE Trans. Electron. Devices*, ED-29, 1982, pp. 1541–7.
68. Yamasaki, K., K. Asai, T. Mizutani and K. Kurumada, 'Self-aligned implantation for n^+ layer technology (SAINT) for high speed GaAs ICs', *Electron. Lett.*, 18, 1982, pp. 119–21.
69. Chen, C. Y., J. Bayruns and N. Scheinberg, 'Reduction of low-frequency noise in a DC-2.5 GHz GaAs amplifier', *IEEE GaAs IC Symposium Technical Digest*, Nashville, TN, November 1988, New York: IEEE, 1988, pp. 289–92.
70. Helix, M. J., S. A. Jumison, C. Chao and M. S. Shur, 'Fan out and speed of GaAs SDFL logic', *IEEE J. Solid State Circuits*, SC-17, 1982, pp. 1226–32.
71. Pedder, D., 'Interconnect packaging of solid state circuits', *European Solid State Circuits Conference*, Manchester, September 1988, *IEEE J. Solid State Circuits*, 24(3), 1979, pp. 698–703.
72. Lee, C. P., R. Zucca and B. M. Welch, 'Orientation effect on planar GaAs Schottky barrier FETs', *Appl. Phys. Lett.*, 37, 1980, pp. 311–13.

73. Asbeck, P. M., C. P. Lee and H. F. Chang, 'Piezo-electric effects in GaAs FETs ...', *IEEE Trans. Electron. Devices*, ED-31, 1984, pp. 1377–80.
74. Greiling, P., R. Lee, H. Winston, A. Hunter, J. Jensen, R. Beaubien and R. Bryan, 'GaAs technology for high speed A/D converters', *IEEE GaAs IC Symposium Technical Digest*, Boston, MA, 23–5 October 1984, New York: IEEE, 1984, pp. 31–3.
75. Lee, C. P., S. J. Lee and B. M. Welch, 'Carrier injection and backgating in GaAs MESFETs', *IEEE Electron. Devices Lett.*, EDL-3, 1982, p. 97.
76. Rocchi, M., 'Status of the surface and bulk effects limiting the performance of GaAs ICs', *Physica*, 129B, 1985, pp. 119–38.
77. Inokuchi, K., M. Tsumotami, T. Ichoka, Y. Sano and K. Kaminishi, 'Suppression of sidegating effects for high performance GaAs ICs', *IEEE GaAs IC Symposium Technical Digest*, Portland, OR, 13–16 October 1987, New York: IEEE, 1987, p. 117.
78. Finchem, E. P., W. A. Vetanen and B. Odekirk, 'Reduction of the backgate effect in GaAs MESFETs by charge confinement at the backgate electrode', *IEEE GaAs IC Symposium Technical Digest*, Nashville, TN, November 1988, New York: IEEE, 1988, pp. 231–4.
79. Solomon, J. E., 'The monolithic op-amp: A tutorial study', *IEEE J. Solid State Circuits*, SC-9, 1974, pp. 314–32.
80. Kraus, J. D. and K. R. Carver, *Electromagnetics*, McGraw-Hill Kogakusha, 1973.
81. Kaupp, H. R., 'Characteristics of microstrip transmission line', *IEEE Trans. Computers*, EC-16, 1967, pp. 185–93.
82. Newett, S. J., D. R. S. Boyd and K. Steptoe, 'CAD tools for GaAs design', *GaAs Technology and Its Impact on Circuits and Systems*, D. G. Haigh and J. Everard (eds), London: Peter Peregrinus, 1989, ch. 4.
83. Williams, J., 'Follow design rules for optimum use of fast comparator IC', *EDN*, June 1985, p. 129.
84. Nishiuchi, K., N. Kobayashi, S. Kuroda, S. Notomi, T. Nimura, M. Abe and M. Kobayashi, 'A sub-nanosecond HEMT 1kb SRAM', *IEEE International Solid State Circuits Conference*, San Francisco, CA, February 1984, Coral Gables, FL: Lewis Winner, 1984, pp. 48–9.
85. Abe, M. and T. Mimura, 'Ultra high-speed HEMT LSI technology', *IEEE GaAs IC Symposium Technical Digest*, New Orleans, LA, October 1990, New York: IEEE, 1990, pp. 127–30.
86. Kroemer, H., 'Theory of a widegap emitter for transistors', *Proc. IRE*, 45, 1957, pp. 1535–7.
87. Kroemer, H., 'Heterostructure bipolar transistors and integrated circuits', *Proc. IEEE*, 70(1), 1982, pp. 13–25.
88. Ho, W. J., M. F. Chang, N. H. Sheng, N. L. Wang, P. M. Asbeck, K. C. Wang, R. B. Nubling, G. J. Sullivan and J. A. Higgins, 'A multifunctional HBT technology', *IEEE GaAs IC Symposium Technical Digest*, New Orleans, LA, October 1990, New York: IEEE, 1990, pp. 67–70.
89. Evans, S., J. Delaney, C. Fuller, D. Boone, C. Dubberley, J. Hoff, J. Stidham, B. de la Torre, M. Vernon and M. Wolowick, 'GaAs HBT LSI/VLSI fabrication technology', *IEEE GaAs IC Symposium Technical Digest*, Portland, OR, 13–16 October 1987, New York: IEEE, 1987, p. 109.
90. Yuan, H., J. B. Delaney, H. D. Shin and L. T. Tran, 'A 4K GaAs bipolar gate array', *IEEE International Solid State Circuits Conference*, Anaheim, CA, February 1986, Coral Gables, FL: Lewis Winner, 1986, pp. 74–5.
91. Kim, M. E., 'GaAs heterojunction bipolar transistor (HBT) device and IC technology for high performance analog/microwave, digital and A/D applications', *High Speed Electronics and Device Scaling Conference*, San Diego, CA, 18–19 March 1990, *Proceedings of the SPIE*, International Society for Optical Engineering, vol. 1288, 1990, pp. 9–20.

92. Hamano, H., T. Ihara, I. Amemiga, T. Futatsugi, K. Ishii and H. Endoh, '20 Gbit/s AlGaAs/GaAs–HBT 2 : 1 selector and decision ICs', *Electron. Lett.*, 27, 1991, pp. 662–4.
93. Wood, T. H., C. A. Burrus, D. A. B. Miller, D. S. Chemla, T. C. Damen, A. C. Gossard and W. Wiegmann, 'High-speed optical modulation with GaAs/GaAlAs quantum wells in a PIN diode structure', *Appl. Phys. Lett.*, 44(1), 1984, pp. 16–18.
94. Whitehead, M., P. Stevens, A. Rivers, G. Parry, J. S. Roberts, P. Mistry, M. Pate and G. Hill, 'Effects of well width on the characteristics of GaAs/AlGaAs multiple quantum well electro-absorption modulators', *Appl. Phys. Lett.*, 53, 1988, p. 956.
95. Whitehead, M., A. Rivers, G. Parry, J. S. Roberts and C. Button, 'A low voltage multiple quantum well reflection modulator with 100 : 1 on : off ratio', *Electron. Lett.*, 25(15), 1989, pp. 984–5.
96. Nobuhara, H., H. Hamaguchi, T. Fujii, O. Aoki, M. Makiuchi and O. Wada, 'Monolithic pin–HEMT receiver for long wavelength optical communications', *Electron. Lett.*, 24, 1988, pp. 1246–8.

Further reading

Soares, R., *Applications of GaAs MESFETs*, Dedham: Artech House, 1983.
Shur, M., *GaAs Devices and Circuits*, New York, NY: Plenum, 1987.
Soares, R., *GaAs MESFET Circuit Design*, Boston, MA: Artech House, 1988.
Mun, J. (ed.), *GaAs Integrated Circuits: Design and Technology*, Oxford: Blackwell Scientific, 1988.
Haigh, D. G. and J. Everard (eds), *GaAs Technology and Its Impact on Circuits and Systems*, London: Peter Peregrinus, 1989.
Milutinovic, V. (ed.), *Microprocessor Design for GaAs Technology*, Englewood-Cliffs, NJ: Prentice Hall, 1990.
Long, S. I. and S. E. Butner, *Gallium Arsenide Digital Integrated Circuit Design*, Wokingham: McGraw-Hill, 1990.
Toumazou, C. and Haigh, D. G., 'GaAs analogue IC design techniques', *Analogue IC Design: The current-mode approach*, C. Toumazou, F. J. Lidgey and D. G. Haigh (eds), London: Peter Peregrinus, 1990, ch. 8.

Index

1/f noise, 10
2DEG, 152

active load, 83, 84, 89
air-bridge, 123
amplifier
  biasing, 80–3
  cascode, 92–9, 100, 101, 107
  differential, 72, 88–92, 94, 145
  double-cascode, 99, 100, 101, 102, 108, 109
  gain, 77, 79, 80, 81, 82, 84, 85, 86, 88, 89, 91, 92, 96, 97, 99, 106, 107, 108, 140, 142, 145, 153
  high gain in, 85
  inverting, 83–4, 101
  offset voltage, 12
  operational, 79, 105–7
  settling time, 109
analog switch, 99–105, 108
analysis, 10, 28, 31, 32, 41, 42, 79, 81, 83
annealing, 117, 121, 122, 144
anodic etch, 124–5, 127
avalanche photo-diode, 157

B_2O_3, 112
backgating, 139–41
bandgap, 1, 8, 130, 152, 153, 158
BFL, 4, 40–2, 43, 45, 46, 47, 54, 55–60, 61–2, 72–3, 75–6, 77–8, 140, 142
biasing, 14, 15, 21, 83, 87, 89, 101, 102, 103, 107
  dc, 79–82
  double-level-shift, 95–9
  process, 144–5
bipolar, 4, 8, 9, 11, 13, 14, 15, 17, 40, 151, 153–5
bond wire, 130
bondpad, 136, 137–8
boric acid, 112
buffer, 46, 48, 55, 64, 69, 73–8, 85–7, 103–4, 115, 117, 142, 145
buffered FET logic, 4, 40–2, 43, 45, 46, 47, 54, 55–60, 61–2, 72–3, 75–6, 77–8, 140, 142
built-in voltage, 21, 71, 152

CAD, 64, 79, 147
capacitance
  decoupling, 131, 148, 149
  gate, 10, 17, 25–6, 27, 29
  input, 93–4, 107, 139
  load, 37, 40, 42, 48, 74, 75, 76, 77, 85, 86, 87, 109
  MIM, 47, 123, 134
  parasitic, 3, 6–7, 101, 103, 109, 123, 135, 136–7, 146, 155–6
  stray, 30, 42
  track, 7, 64, 70, 136–7
capacitance to ground, 7, 136–7

capacitor, 26, 32, 46, 47, 48, 79, 81, 94, 97, 99, 105, 107, 108, 109, 123, 124, 134, 148, 149
capacitor-coupled logic, 46–7
capacitor-diode FET logic, 47–8
cascode
  amplifier, 92–9, 100, 101, 102
  source-follower, 93–4, 97–9
  switch-driver, 105–6, 108–9
CCL, 46–7
CDFL, 47–8
characteristic impedance, 9, 146–7, 149
circuit analysis, 28, 31–3, 79–80
clock feedthrough, 103–5, 109
common-mode rejection ratio, 88, 90
compensation capacitor, 107
complementary JFETs, 9, 11
complex gates, 54, 65, 152
conductance frequency dependence, 28
contact
  ohmic, 22, 30, 122, 126, 134, 139, 140, 153
  resistance, 18, 27, 121, 122, 126, 134, 152, 153
  Schottky, 1, 18, 20–1, 32–3, 35, 47, 71, 87–8, 117–18, 122, 124, 127–8, 134, 139, 142, 143, 152, 153, 154
cost, 1, 2, 4, 11, 24, 30, 108, 119
cross-coupled load, 97, 98, 101
cross-coupling, 95–7
cross-talk, 138, 155, 156
crystal growth, 110–14
current-limiting, 71
current-steering, 43–4, 55, 62
Curtice, W.R., 31

DCFL, 4, 21, 37–9, 40, 42, 43, 46, 48, 54, 55, 56, 59, 60, 61, 62, 71, 72, 74, 75, 76, 77, 107, 135, 144
decoupling capacitor, 131, 148, 149
depletion layer, 6–7, 14, 15, 17–18, 19, 20, 21–4, 25, 26, 27, 35, 124, 125, 139, 151, 158
depletion-mode, 22–3, 36, 40, 83, 88, 100, 107, 117, 133, 144, 152
design techniques, 1, 61–4, 132, 155, 166
dielectric, 1, 2, 14, 122, 123–4, 134
differential amplifier, 72, 82, 88–92, 94, 145
diode, 18, 20–1, 35, 37, 39, 40, 41–2, 43, 45, 46–8, 49, 55, 59, 71, 72, 83, 86, 87–8, 90, 102, 103, 105, 109, 117, 124–5, 134, 157
  layout, 142–3
  model, 32–3
direct-coupled FET logic, 4, 21, 37–9, 40, 42, 43, 46, 48, 54, 55, 56, 59, 60, 61, 62, 71, 72, 74, 75, 76, 77, 107, 135, 144
dislocation density, 110, 112, 113
dislocation-free, 113
dissociation, 110
distortion, 80, 100, 145, 146
doping level, 17, 20, 115, 124–5, 151, 153, 155
double cascode, 99, 100, 101, 102, 108, 109
double heterojunction bipolar transistor, 153
double-level-shift biasing, 95–9, 101, 102
driver, 74–7, 100, 101, 103–6, 107, 108, 109
dynamic range, 80, 107, 108

ECL, 4, 38, 43, 71–7
electron mobility, 1, 2, 4, 6–7, 11, 15, 18, 24, 29, 99, 112, 151
electron velocity, 19–20, 30, 141
emitter-coupled logic, 43, 71
energy states, 2, 8, 10, 12, 16, 17, 20, 28, 123, 143
enhancement-mode, 22–3, 36, 37, 40, 83, 100, 117, 133, 152
equivalent circuit, 72, 83, 84, 85
error function, 53
etch, 119, 120, 126, 128, 130, 144
  anodic, 124–5, 127

fan-in, 59–60, 64, 69, 70
fan-out, 59–60, 64, 69, 70
feedback, 45, 69, 80, 81, 82, 107, 145
feedforward, 107
flash annealing, 144
flicker noise, 10
fragility, 2, 12, 131
full-adder, 61–4, 66, 67

gain, 77, 79, 80, 81, 82, 84, 85, 86, 88, 89, 91, 92, 96, 97, 99, 106, 107, 108, 140, 142, 145, 153
gate arrays, 4–5, 48
gate capacitance, 10, 17, 25–6, 27, 29
gate orientation, 138–9
gate parasitic resistance, 27–8
gate-priority, 127, 128
g_m, 25, 26, 77, 80, 84, 86, 92, 99, 126, 139, 143
g_o, 24–5, 28–9, 31, 32, 80, 84, 85, 86, 90, 92, 97, 99, 105, 143
ground connection, 130–1, 149–50
ground plane, 147, 149–50

HB, 110–12
heterojunction, 15, 117, 151, 153, 154
heterojunction bipolar transistor, 15, 151, 153–5
HJBT, 15, 151, 153–5
hole mobility, 1, 2, 11, 15, 17, 153
horizontal Bridgman, 110–12
hybrid-pi, 25, 31, 79–80, 83

I-V characteristics, 20, 21, 23, 25, 36
I_{DSS}, 24, 45, 71
IGFET, 8, 12, 15–16, 17, 143
inductance, 130, 135, 136, 138, 146, 148, 149, 150, 155
input buffer, 71–3, 145
input capacitance, 74, 77, 93, 94, 107, 139
input impedance, 45, 71, 79, 80, 84, 86, 88, 90, 102
insulated-gate field effect transistor, 8, 12, 15–16, 17, 143
interconnect, 122–3, 124, 134, 136, 158
inversion layer, 15, 16, 17
inverter, 34–46, 55, 59, 72, 74, 77, 80–1, 83–4, 91, 92, 93–4, 95, 96, 97, 98, 99, 100, 102, 103, 104, 105, 106, 107, 108, 109
ion implantation, 26, 114, 115–17
I_S, 21, 32

JFET, 9, 11, 16–18, 25, 31–2
junction field effect transistor, 9, 11, 16–18, 25, 31–2

lattice mismatch, 10, 156
layout, 12, 64, 132–45
  diode, 142–3
  MESFET, 138–40, 143–4
  parasitics, 64, 135–7
  process tolerance, 144–5
  resistor, 141
  rules, 132–4
  temperature gradients, 12, 145
LEC, 110, 112–14
level of integration, 3, 35, 54, 55, 61
level-shift, 40–2, 45, 47, 55, 59, 64, 75, 95–9, 101, 102, 107, 142, 144–5
lift-off, 119–20, 144
linear region, 24, 38, 80
liquid encapsulated Czochralski, 110, 112–14
load
  capacitance, 37, 40, 42, 48, 74, 75, 76, 77, 85, 86, 87, 109
  cross-coupled, 97, 101
  effects in switch-drivers, 103–5
  passive *vs* active, 34–7, 83, 84, 88–90
  width ratio in DCFL, 38–40
load-line, 36–7
logic family comparison, 42–3, 46, 54–5, 59, 61–2
logic high, 34, 37, 39, 41, 49, 50, 55, 59, 60, 71, 72, 74, 75, 76, 135, 138
logic low, 34–5, 37–8, 41, 49, 55, 59, 69, 76, 135, 138
logic swing, 21, 54
LSI/VLSI, 35, 54, 55, 61, 112, 113, 122, 124, 139, 144, 152, 155

MBE, 114, 115–17, 152, 153
memories, 8–9
mesa, 120
metal-organic chemical vapour deposition, 114, 115, 116, 117, 152, 153
Miller effect, 77, 94, 107
MIM capacitor, 47, 123, 134
mobility, 1, 2–4, 6, 7, 11, 14, 15, 17, 18, 24, 29, 99, 112, 151–2, 153
MOCVD, 114, 115, 116, 117, 152, 153
model, 25, 28–9, 31–3, 79–80, 83
modulator, 157

molecular beam epitaxy, 114, 115–17, 152, 153
MQW, 157
multi-quantum well, 157
mushroom gate, 27

N^+ priority, 127, 129
NAND, 55–60, 61, 62
noise
  1/f, 10
  immunity, 73
  low-noise amplifiers, 157
  margin, 21, 39, 41, 42, 49–55, 59–60, 69, 76, 84, 135, 138, 142, 144, 148, 152
  removal, 88–90
  thermal, 10–11, 27
NOR, 55–60, 61, 62, 144

offset voltage, 12, 84, 86, 88, 101, 102–3, 140, 144
ohmic contact, 22, 30, 122, 126, 134, 139, 140, 153
on-resistance, 34, 37, 49, 55, 59, 100, 101
open-drain, 76, 77
open-source, 74, 75, 76
operational amplifier, 79, 105–7
opto-electronic, 2, 155–8
output buffer, 64, 69, 145
output conductance, 24–5, 28–9, 31, 32, 80, 84, 85, 86, 90, 92, 97, 99, 105, 143

packaging, 7, 9, 53, 74, 128–31, 138, 155, 156
packing density, 35, 39, 55, 140, 145
parasitic
  capacitance, 3, 6–7, 101, 103–5, 109, 123, 135–7, 146, 155–6
  inductance, 130, 135–6, 138, 146, 149
  IGFET, 143–4
  resistance, 3, 8, 10, 11, 14–15, 25, 26–8, 46, 55, 117, 121, 122, 123, 125, 126–9, 134, 135–7, 139, 142–3, 146, 152, 153, 155–6
passivation, 123–4
passive load, 34–7, 83–4, 88–90
PBN, 112
PECVD, 124
photolithography, 11, 12, 53, 110, 117, 118–20, 127
photoresist, 118–20, 123, 124, 127, 128, 130, 144
pinch-off, 24
planar, 117, 119, 121, 126, 127, 139, 153
plasma enhanced chemical vapour deposition, 124
polyimide, 124
power dissipation, 2–5, 9, 29, 37, 41, 42, 43, 45, 46, 47, 48, 54, 61, 62, 69, 72, 76, 96, 99, 103, 130, 145, 155, 156, 157
power supply rails
  analog, 83, 101
  BFL, 40–1
  DCFL, 71
  parasitics, 130, 135–6, 138, 148–9
  SCFL, 46
  SDFL, 43
power-delay product, 3–4, 29–30
process bias, 144
process flow, 117–18
process tolerance, 42, 43, 144–5
propagation delay, 2–5, 8, 29–30, 42, 43, 62, 64–70, 139, 155
properties (GaAs *vs* Si), 1–2
push-pull, 48, 74–5, 77, 97, 107
pyrolytic boron nitride, 112

radiation hardness, 2, 7–9, 18
  soft errors, 11
  total dose, 8
  transient, 8, 123
ratioed-logic, 34–43, 46, 56–60, 61–2
recessed gate, 26, 126–7
resistivity, 1, 2, 6, 11, 16, 18, 30, 35, 110, 112, 113, 114, 117, 135, 143, 144, 152
resistor
  characteristics, 18–20
  feedback network, 80–4
  layout, 141, 143
  load in inverters, 34–7
  saturation, 71
  terminating, 72–8, 146, 149
ring oscillator, 69–70

SAINT, 28, 127, 129
saturated resistor, 20, 37, 71
saturation
  current, 21, 37
  region, 23–5, 31, 38, 41, 80–1, 83, 92, 93, 96, 99, 107, 108, 145
  velocity, 1, 19, 20, 141
SBFL, 4, 48–9, 55, 56
scaling (of dimensions), 29–30
SCF, 108–9
SCFL, 43–6, 54–5, 57, 58, 59, 61, 62–8, 103–4
Schottky contact
  barrier height, 1, 152
  capacitor, 47
  characteristics, 18, 20–1
  diode model, 32–3
  forward voltage drop, 35
  in HJBT, 153, 154
  layout, 139, 142, 143
  level-shift, 87–8
  processing, 117–8, 122, 134
  self-aligned gate, 127–8
  static protection, 71
Schottky diode FET logic, 42, 43, 45, 46, 47, 54, 55, 56, 57, 59, 61, 62, 137, 140
SDFL, 42, 43, 45, 46, 47, 54, 55, 56, 57, 59, 61, 62, 137, 140
self-aligned implantation for N^+ layer technology, 28, 127, 129
self-alignment, 26, 121, 126, 127–9, 132, 133, 135, 139
semi-insulating substrate, 2, 5–7, 120, 123, 124, 125, 136
settling time, 109
short-channel effect, 30, 121
sidegating, 139–41
silicon nitride, 16, 121, 124, 127, 128
simulation, 31–3, 61, 64, 79
soft-errors, 11
source-coupled FET logic, 43–6, 54–5, 57, 58, 59, 61, 62–8, 103–4
source-follower, 40–2, 48, 75–6, 77, 85–8, 91, 107
  cascode, 92–4
  double-cascode, 97–9, 102, 109
  layout, 140, 141
  switch-driver, 101, 102, 103
speed, 2–4, 7, 9, 11, 14, 15, 17, 18, 24, 25, 30, 31, 36, 37, 40, 44, 46, 48, 58, 64, 69, 71, 75, 76, 105, 108, 122, 126, 127, 128, 130, 137, 138, 139, 146, 148, 149, 151, 152, 153, 155, 156
SPICE, 31–2, 39, 61–2, 64, 79
strain, 10, 152, 156
substrate
  backgating effects, 139, 140
  currents, 30, 131
  growth, 110–14
  radiation effects, 8
  semi-insulating properties, 2, 5–7, 120, 136, 137
  thermal properties, 12
super buffer FET logic, 4, 48–9, 55, 56
switch-driver, 99–105, 106, 108, 109
switched-capacitor filter, 108–9

terminating resistor, 72–8, 146, 149
thermal conductivity, 1, 2, 12, 128, 145
thermal gradients
  in processing, 112, 144
  on IC, 12, 145
thermal noise, 10–11
thershold voltage, 8, 16, 22, 24, 30, 37, 38, 39, 40, 41, 42, 49, 50, 51, 52, 54, 55, 61, 72, 83, 84, 86, 87, 92, 96, 100, 101, 107, 113, 114, 124, 125, 126, 127, 139, 140, 144, 155
tolerance
  alignment, 132
  process, 42, 43, 144, 145
total dose immunity, 8
track
  capacitance, 7, 64, 70, 136–7
  parasitics, 135–7, 143–4, 146–8, 149
  processing, 122–3
track–ground capacitance, 7, 136–7
track–track capacitance, 7, 136–7
transconductance, 25, 26, 77, 80, 84, 86, 92, 99, 126, 139, 143
transfer characteristics, 49–52, 59, 60
transfer efficiency, 86–7, 91–2
transient radiation, 8, 123
transistor–transistor logic, 4, 8, 71, 73–4, 76–8
transmission lines, 71, 146–8, 149, 155

TTL, 4, 8, 71, 73–4, 76–8
two-dimensional electron gas, 152

uniformity, 54, 61, 112, 114, 116, 117, 127, 139, 144

vapour-phase epitaxy, 114, 115, 117
V_{bi}, 21, 71, 152
velocity saturation, 1, 19, 20, 141
via-holes, 122–3, 124
VLSI/LSI, 35, 54, 55, 61, 112, 113, 122, 124, 139, 144, 152, 155
V_P, 24
VPE, 114, 115, 117
V_T, 8, 16, 22, 24, 30, 37, 38, 39, 40, 41, 42, 49, 50, 51, 52, 54, 55, 61, 72, 83, 84, 86, 87, 92, 96, 100, 101, 107, 113, 114, 124, 125, 126, 127, 139, 140, 144, 155

yield, 4, 24, 30, 31, 51–5, 110, 113, 121, 132, 135, 144